Statistics for Social Science and Public Policy

Advisors:
S. E. Fienberg   D. Lievesley   J. Rolph

Springer Science+Business Media, LLC

Statistics for Social Science and Public Policy

Valen E. Johnson
James H. Albert

Ordinal Data Modeling

With 73 illustrations

 Springer

Valen E. Johnson
Institute of Statistics
  and Decision Sciences
Duke University
Durham, NC 27706
USA

James H. Albert
Department of Mathematics
  and Statistics
Bowling Green State University
Bowling Green, OH 43403
USA

Advisors

Stephen E. Fienberg
Department of Statistics
Carnegie Mellon University
Pittsburgh, PA 15213
USA

Denise Lievesley
Institute for Statistics
Room H.113
UNESCO
7 Place de Fontenoy
75352 Paris 07 SP
France

John Rolph
Department of Information and
  Operations Management
Graduate School of Business
University of Southern California
Los Angeles, CA 90089
USA

Library of Congress Cataloging-in-Publication Data
Johnson, Valen E.
        Ordinal data modeling / Valen E. Johnson, James H. Albert.
             p.   cm. — (Statistics for social science and public policy)
        Includes bibliographical references and index.
        ISBN 978-1-4757-7290-6   ISBN 978-0-387-22702-3 (eBook)
        DOI 10.1007/978-0-387-22702-3

        1. Social sciences—Statistical methods.   2. Policy sciences—
Statistical methods.   3. Numbers, Ordinal.   I. Albert, James H.
II. Title.   III. Series.
HA29.J588   1999
519.5—dc21   98-51801

Printed on acid-free paper.

© 1999 Springer Science+Business Media New York
Originally published by Springer-Verlag New York, Inc. in 1999
Softcover reprint of the hardcover 1st edition 1999

Production managed by Jenny Wolkowicki; manufacturing supervised by Joe Quatela.
Typeset by The Bartlett Press, Inc., Marietta, GA.

9 8 7 6 5 4 3 2 (Corrected second printing, 2000)

ISBN 978-1-4757-7290-6   SPIN 10786585

Preface

This book was written for graduate students and researchers in statistics and the social sciences. Our intent in writing the book was to bridge the gap between recent theoretical developments in statistics and the application of these methods to ordinal data. Ordinal data are the most common form of data acquired in the social sciences, but the analysis of such data is generally performed without regard to their ordinal nature.

The first two chapters of the book provide an introduction to the principles and practice of Bayesian inference, and the modern computational techniques that have made Bayesian inference possible in the models that are described later in the text. Together with a previous course in introductory statistics and a knowledge of multiple regression, these two chapters provide the background necessary for studying the remainder of the book. Readers familiar with this material can safely skip to the third chapter, where models for binary data are introduced.

The third chapter describes binomial regression models from both classical and Bayesian perspectives. Considerable attention is given to residual analyses and goodness-of-fit criteria. Latent variable interpretations of binomial regression models are also emphasized.

Chapter 4 extends the results of the previous chapter to regression models for ordered categorical data, or ordinal data. Motivation for these models is again drawn from a latent variable interpretation of categorical data, thus exposing the close connections between the underlying model structures and computational algorithms common to both binomial and ordinal regression models. As in Chapter 3, considerable attention is paid to residual analyses from both classical and Bayesian perspectives.

Multirater ordinal data and regression models are introduced in Chapter 5. To effectively handle the complications introduced by the implicit rater variability associated with such data, a Bayesian inferential structure is assumed throughout. The multirater ordinal regression models developed in this chapter provide a general framework for a large number of psychometric models, including the item response models of Chapter 6 and the graded response models in Chapter 7. ROC analysis is discussed within the general context of multirater ordinal regression data in the latter sections of this chapter.

Chapters 6 and 7 describe item response and graded response models. Item response models are special cases of multirater ordinal data models in which responses are binary. Basic terminology associated with item response models is introduced, and a hierarchical Bayesian model that represents a compromise between one-parameter Rasch-type models and more general two-parameter models is proposed. Numerous diagnostic plots for item response models are illustrated. Chapter 7 closes with a brief description of graded response models and illustrates the principles of these models in a case study of undergraduate grade data.

Much of the data described in the book and many of the computational algorithms used in the analyses are available electronically from the authors or Mathworks. On the world-wide web, this material may be obtained from

www-math.bgsu.edu/˜albert/ord_book

or

ftp://ftp.mathworks.com/pub/books/johnson.

We would like to thank several individuals for their assistance in preparing the manuscript. David Draper provided very helpful comments on an early draft. John Kimmel, our editor, was also most helpful in bringing closure to the process. We would also like to thank a number of students who contributed to proofreading: Sining Chen, Maria De Iorio, Courtney Johnson, Dae Young Kim, Scott Lynch, Randy Walsh, and Hongjun Wang. We are also grateful to Dr. Robert Terry and Brian Skotko for their permission to use previously unpublished data.

Finally, we thank our wives, Pam and Anne, and our children Kelsey and Emily, and Lynne, Bethany, and Steven for encouragement and inspiration.

Duke University   Valen E. Johnson
Bowling Green State University   James H. Albert

December 9, 1998

Contents

1
Review of Classical and Bayesian Inference

In this chapter, we review basic principles of classical and Bayesian inference. We begin by examining procedures that can be used to estimate a binomial proportion, then proceed to inference regarding the mean and variance of a normal population. Following this, we return to binomial data and illustrate the use of hierarchical Bayesian models in estimating sets of parameters thought to have similar values.

1.1 Learning about a binomial proportion

In the fall of 1997, the faculty and administration of Duke University decided to review and revise the university's curriculum. Periodic curriculum reviews are common at most universities and often focus on the number and type of courses undergraduates must take in order to receive their baccalaureate degree. Duke had not undergone a major curriculum review in several years, and with the recent appointment of three new deans in its college of arts and sciences, the time was clearly ripe.

Curriculum reviews may be initiated for any number of reasons. At Duke, there seemed to be a perception among many faculty that academic standards were declining. Evidence cited for this decline included rapidly increasing grade distributions, surveys that indicated that many students were spending less time studying, and a decreasing trend in the number of upper-level undergraduate courses taken by the average student. To reverse these perceived trends, several remedies were proposed almost as soon as the curriculum review committee was established. Foremost among these was a proposal to increase the number of distributional requirements imposed on undergraduates.

Essentially, distributional requirements are general education requirements. Duke undergraduates are required to take courses in five of six major fields of study and must take "100-level" courses in four of these. Increasing the number of distributional requirements was one mechanism through which many faculty thought academic standards could be reestablished.

Or could they?

To investigate this question, the authors conducted a survey in one of the larger statistics classes offered at Duke. In the survey, students were asked to answer questions about the course they were scheduled to take immediately following their statistics class. Among the questions asked about this class, which differed from one student to the next, was whether the class was taken to satisfy a distributional requirement, a major requirement, or an elective. Students were also asked whether their prior knowledge of how the course would be graded affected their decision to enroll in it. One hypothesis that we wished to investigate using this survey data was whether "grade shopping" was more common in courses taken to satisfy distributional requirements than it was in other classes. If so, increasing the number of distributional requirements might actually have an effect opposite to that intended by the curriculum review committee; that is, more easy classes with high grade distributions might be taken if such a change were enacted, because more students might register for such classes to satisfy what they perceived to be unnecessary and irrelevant distributional requirements.

Of the students completing the survey, 16 reported that the class they were scheduled to take immediately following their statistics class was taken to satisfy a distributional requirement. Of these, 5 reported that their prior knowledge of the grading policy in the class had no effect on their decision to enroll in it; the remaining 11 reported that the anticipated grading policy had an effect that ranged between marginal to very significant.

To facilitate a more quantitative discussion of these issues, let us for the time restrict our attention to estimating the proportion p of courses taken to satisfy distributional requirements in which the grading policy was reported to play no part in a student's decision to enroll. Of course, to determine this proportion exactly, we would have to ask every student enrolled at Duke University in 1997 whether grading policy had influenced their decision to enroll in each distributional requirement course they had taken. This would be a very laborious process, and it is unlikely that responses could be obtained from all students.

Alternatively, we can estimate this proportion using the survey data collected. To do so, we might assume that the students questioned represented a random sample of all students at Duke University and that the distributional requirements taken by the sixteen students in the class represented a random sample of all courses taken by all students to satisfy a distributional requirement. By random sample, we mean that each individual or course had an equal chance of being included in our survey. Of course, this assumption is probably not exactly true in this survey, but because there does not seem to be any obvious reason to think that the responses to the survey are related to inequities in inclusion probabilities, we will proceed as if the sampled population might be treated as a random sample from the overall population of students taking distributional requirements.

For each of these 16 students and the distributional requirements they were taking, we can define a binary random variable indicating whether or not the anticipated grading policy had influenced the student's decision to enroll. In general, a binary (or Bernoulli) random variable is a random variable that can assume one of two values. Based on the value of a binary response for every individual, an entire population can be divided into two groups, and then summarized by the proportion of individuals falling into each group.

1.1.1 Sampling: The binomial distribution

In conducting the survey, a sample of size $n = 16$ students taking distributional requirements was obtained, and of this sample, $y = 5$ students reported that grading played no part in their enrollment decision. Assuming that this value of n is small compared to the actual number of students in the population of interest,[1] and letting p denote the probability that a student selected randomly from this population reports that grading policy played no part in their enrollment decision, it follows that y has a binomial distribution with denominator n and success probability p. The probability mass function for y is

$$f(y|p) = \binom{n}{y} p^y (1-p)^{n-y}, \qquad y = 0, 1, ..., n, \qquad (1.1)$$

where $\binom{n}{y}$ is the binomial coefficient $\frac{n!}{y!(n-y)!}$.

More generally, (1.1) describes the probability of observing y "successes" out of n items "tested" when items (persons) are selected randomly from an infinite population in which the proportion of items labeled as success is p.

1.1.2 The likelihood function

Equation (1.1) describes the probability of observing y successes out of n trials when the success probability p is known; we denote this *sampling density* by $f(y|p)$, the sampling density of y given p. After completing the survey, however, the value of y is known, and we are actually interested in estimating the success probability p. One general way for obtaining an estimate of p (or other parameters appearing in the sampling density) is based on maximizing the sampling density

---

[1] This assumption is made for two reasons. First, we wish to ignore the effects of having a (small) finite population, so that the proportion of students surveyed is small compared to the total population. Otherwise, we would have to account for the fact that our knowledge of p in the sampled population becomes exact as the proportion of the population sampled approaches one. Second, we wish to ignore the differences between sampling with replacement and without—in an "infinite" population, the chance of sampling the same item twice is zero.

when it is regarded as a function not of y, but of the unknown parameter p. When regarded as a function of p, the sampling density (1.1) is called the *likelihood function*.

To make our discussion of maximum likelihood estimation more general, let us suppose that our parameter of interest is θ (in the survey example, $\theta = p$) and that the sampling distribution is $f(y|\theta)$. After observing data y, the *likelihood function* is defined to be the probability mass function or probability density function treated as a function θ, rather than as a function of y. Given y, the likelihood function can be computed over a range of values for the parameter θ, and relative values of the likelihood function provide an indication of which parameter values are most consistent with the observed data. Intuitively, it makes sense to estimate the unknown parameter θ by the value which makes the observed data most likely. The value of θ at which the likelihood function achieves its maximum is called the maximum likelihood estimate, or MLE.

To illustrate, recall that 16 students were sampled in the class survey and 5 students reported no grade effect. The probability mass function, for a known value of p, is then the probability of observing 5 out of 16 responding in this way. The likelihood, denoted by $L(p)$, can thus be written

$$L(p) = \binom{16}{5} p^5 (1-p)^{11}, \qquad 0 < p < 1. \qquad (1.2)$$

This function can be computed for values of p between 0 and 1. If $p = .2$, then the probability of getting 5 in agreement out of 16 is $L(.2) = \binom{16}{5}(.2)^5(1-.2)^{11} = .120$. If $p = .4$, the probability of getting 5 agreements out of 16 is $L(.4) = .162$. Table 1.1 summarizes the likelihood function for equally spaced values of p between .1 and .9.

If the probability of 5 agreements out of 16 is computed for all proportions between 0 and 1, we obtain the likelihood function $L(p)$, displayed in Figure 1.1.

1.1.3   Maximum likelihood estimation

We can see from Table 1.1 and Figure 1.1 that some values of the population proportion p are more consistent with our sample result than are other values. For example, the probability of five agreements in our sample when $p = .1$ is $L(.1) = .014$; when $p = .2$, this probability is $L(.2) = .120$. Thus, a population in which 20% of students agree with the statement regarding the influence of grade policy on course enrollment is more consistent with our data than a population in which 10% of students agree.

p	.1	.2	.3	.4	.5	.6	.7	.8	.9
$L(p)$	.014	.120	.210	.162	.067	.014	.001	.000	.000

TABLE 1.1. Values of the likelihood for a sample result of 5 in agreement out of 16 respondents for survey example.

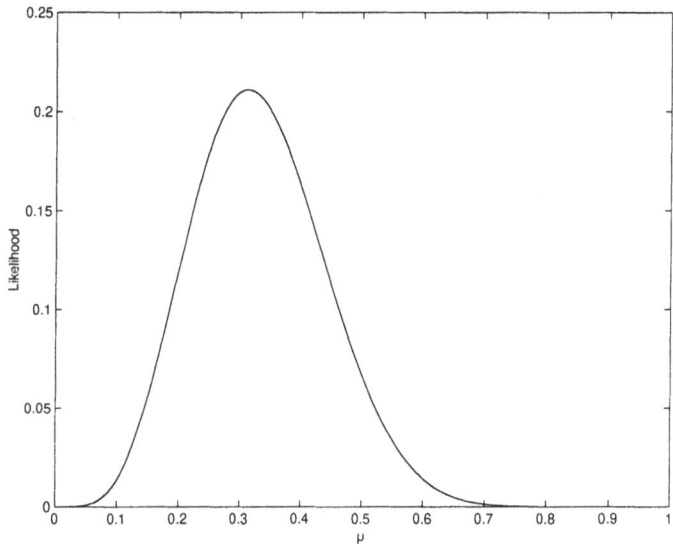

FIGURE 1.1. The likelihood function $L(p)$ for a binomial sample of 5 agreements out of 16.

To find the MLE, we maximize the likelihood function $L(p)$ as a function of p. Equivalently, we can find the value of p which maximizes the logarithm of the likelihood,[2] $l(p) = \log L(p)$. In the general case in which we observe y successes in n trials, the log-likelihood is given by

$$l(p) = \log \binom{n}{y} + y\log(p) + (n - y)\log(1 - p), \qquad 0 < p < 1.$$

Taking the first derivative of $l(p)$ with respect to p and setting the resulting expression equal to 0, we see that the MLE for p must satisfy

$$l^{(1)}(p) = \frac{y}{p} - \frac{n - y}{1 - p} = 0,$$

where $l^{(1)}(p)$ is the derivative of the log-likelihood. Solving this equation for p, we find that the MLE is

$$\hat{p} = \frac{y}{n}.$$

---

[2] When finding the MLE, the natural logarithm of the likelihood function is preferable to the likelihood function for several reasons. First, the log-likelihood increases (or decreases) whenever the likelihood increases (decreases), so the value of the parameter that maximizes one also maximizes the other. Second, differentiating a product is analytically difficult, while differentiating a sum is easy. Finally, the curvature of the log-likelihood is related to the sample variance of the MLE, as we illustrate later in this section.

1.1.4   The sampling distribution of the MLE

Classical inference concerning the value of a parameter is based on the sampling distribution of an estimator. This sampling distribution describes the variation of the estimator under repeated sampling from the given population. In the binomial setting, the sampling distribution of $\hat{p}$ is the probability distribution of this estimate if repeated binomial samples are taken with sample size n and fixed probability of success p. In some cases, sampling distributions of estimators are known exactly. For example, the sample mean of n draws from a normal population with mean μ and standard deviation σ is known to have a normal distribution with mean μ and standard deviation $\sigma/\sqrt{n}$. Unfortunately, in many situations, the sampling distribution of an estimator cannot be derived analytically. In such circumstances, classical inference often relies on asymptotic results. These results provide an approximation to the sampling distribution of the estimator when the sample size is "large." In the case of the MLE, there is a convenient asymptotic approximation to the sampling distribution that pertains in most applied settings. For large n, the MLE of a scalar parameter, say $\hat{\theta}$, is approximately normally distributed with mean θ and standard deviation $(-l^{(2)}(\hat{\theta}))^{-1/2}$, where $-l^{(2)}(\hat{\theta})$ denotes the negative of the second derivative of the log-likelihood evaluated at the MLE.

In the case of a binomial proportion, the second derivative of the log-likelihood is

$$l^{(2)}(p) = -\frac{y}{p^2} - \frac{n-y}{(1-p)^2}.$$

Substituting the MLE $\hat{p}$ into this expression and taking a reciprocal and a square root, we obtain

$$(-l^{(2)}(\hat{p}))^{-1/2} = \sqrt{\frac{\hat{p}(1-\hat{p})}{n}}.$$

Thus, for binomial data, the MLE $\hat{p}$ is asymptotically normally distributed with mean p and standard deviation $\sqrt{\hat{p}(1-\hat{p})/n}$. The standard deviation of an estimator is called its standard error (se). Using this normal approximation, the approximate standard error of the MLE is $\sqrt{\hat{p}(1-\hat{p})/n}$.

1.1.5   Classical point and interval estimation for a proportion

From a classical viewpoint, a point estimate of the proportion of students in agreement with the survey question is provided by the MLE, given by

$$\hat{p} = \frac{y}{n} = \frac{5}{16} = .312.$$

The approximate standard error for this estimate is

$$se(\hat{p}) = \sqrt{\frac{\hat{p}(1-\hat{p})}{n}} = \sqrt{\frac{.312(1-.312)}{16}} = .116.$$

An interval estimate for the population proportion p can be obtained using the asymptotic normal sampling distribution of the MLE $\hat{p}$. In this case, the standardized statistic

$$\frac{\hat{p} - p}{\text{se}(\hat{p})} = \frac{\hat{p} - p}{\sqrt{\frac{\hat{p}(1-\hat{p})}{n}}}$$

is approximately distributed as a standard normal deviate. It follows that a γ-level confidence interval for p is given by

$$(\hat{p} - z_{\gamma/2}\,\text{se}(\hat{p}), \hat{p} + z_{\gamma/2}\,\text{se}(\hat{p})),$$

where $z_{\gamma/2}$ is the $1 - \gamma/2$ quantile of the standard normal distribution. In this example, if $\gamma = .90$, then a 90% confidence interval for p is

$$(.312 - 1.645 \times .116, .312 + 1.645 \times .116) = (.121, .503).$$

The "confidence" of this interval is a reflection of the initial probability statement about the sampling distribution of $\hat{p}$. In repeated sampling, one expects the random interval $(\hat{p} - z_{\gamma/2}\,\text{se}(\hat{p}), \hat{p} + z_{\gamma/2}\,\text{se}(\hat{p}))$ to include the unknown parameter p with probability close to γ.

1.1.6 Bayesian inference

Bayesian inference is based on the subjective view of probability. Many events in our world are uncertain, such as next week's weather, a person's health 10 years in the future, or the proportion of the population of Duke students who report that grading policies are inconsequential in their selection of distributional requirements. A probability can be viewed as a measure of a person's opinion about the truth of an event. This type of probability is subjective. For example, a person's assessment of the probability that a Duke student does not consider the grading policy when selecting distributional requirements depends on his or her personal belief about the opinion of this population. Different people may have different beliefs about this proportion depending on their backgrounds. In addition, the subjective probability of an event is conditional—it depends on the current state of knowledge of the individual. As one obtains new information, one's probabilities about events change. In the survey example, one's assessment of the probability of no grading effect may change as results of the survey are revealed. Bayes' rule is the recipe for updating one's probabilities about uncertain events when new information is obtained.

1.1.7 The prior density

The parameter of interest in the undergraduate student survey is the value of the proportion p. Conceivably, p could be any value in the interval $(0, 1)$. Within the Bayesian paradigm, a person expresses one's belief about the location of this population proportion, before observing any data, by means of a probability density

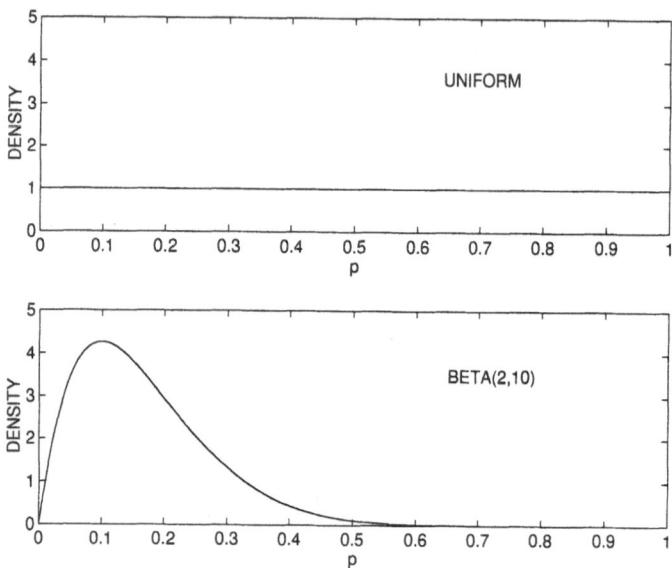

FIGURE 1.2. Two prior densities for a proportion. The top figure is a uniform density that reflects vague prior beliefs; the bottom figure represents an informative density which reflects the prior belief that the proportion is small.

on the unit interval. This probability density is usually called a *prior density*, since it reflects a person's beliefs *prior* to observing survey data.

To illustrate the use of a prior density in inference, consider an individual who has little knowledge about this proportion. For this individual, all values of the proportion p between 0 and 1 may be equally plausible a priori. To model this information about p, this individual's prior beliefs may be summarized by a uniform density on the unit interval; that is,

$$g(p) = 1, \qquad 0 < p < 1,$$

as graphed in Figure 1.2 (top). This density might be summarized by computing probabilities—or areas—under the density. For example, $\Pr(p < .25) = \Pr(p > .75) = .25$. The uniform density reflects the prior belief that the unknown proportion is as likely to be small (say less than .25) as large (greater than .75). This prior is often called vague or noninformative, since it reflects a lack of prior information about the value of the proportion.

A second person may have more precise beliefs about the location of the proportion's value. Suppose this individual believes that only a small proportion of students would think that grading played no role in their decision to enroll in a class to satisfy a distributional requirement. For this person, the prior density for p might be concentrated on small values in the unit interval. One density that might represent this prior information is displayed in Figure 1.2 (bottom). Note that most of the mass of this density lies between 0 and .5. The probabilities that p falls in the intervals (0, .25) and (.75, 1) for this particular prior density are .80 and approximately 0, respectively. This distribution reflects an individual's opinion that the proportion of students agreeing with this statement is likely to be under 25%. Such a prior is called an informative prior.

1.1.8   Updating one's prior beliefs

The prior density $g(p)$ reflects the beliefs of a researcher before any data are observed. Once data are obtained, the prior density is updated on the basis of the new information. We call the updated probability distribution on the parameter of interest the *posterior distribution*, since it reflects probability beliefs posterior to, or after, seeing the data.

According to Bayes' theorem, the posterior probability distribution is computed (up to a proportionality constant) by multiplying the likelihood function by the prior density. Symbolically, the posterior density is obtained according to the simple updating strategy

$$\text{POSTERIOR} \propto \text{PRIOR} \times \text{LIKELIHOOD},$$

where $\propto$ denotes a proportionality relationship. In terms of probability density functions,

$$g(p|\text{data}) \propto g(p)L(p).$$

Continuing with the survey example above, assume that a random sample of size $n = 16$ was selected from the population of Duke students taking a distributional requirement and that $y = 5$ students reported no effect of grading policy on their decision to enroll. As defined above, the likelihood function is the value of the probability density function for the observed data, given the value of the data. From our earlier discussion of the classical analysis, the likelihood function for the survey data is given by

$$L(p) = \binom{n}{y} p^y (1 - p)^{n-y}, \qquad 0 < p < 1.$$

Combining the prior density and the likelihood, it follows that the posterior density is

$$g(p|\text{data}) \propto g(p)p^y (1 - p)^{n-y}, \qquad 0 < p < 1,$$

where $g(p)$ denotes the particular prior density chosen by the investigator to represent any prior knowledge available for p. If the prior density is described by the uniform density of Figure 1.2 (top), then the posterior density of p is proportional to

$$g(p|\text{data}) \propto p^5 (1 - p)^{11}, \qquad 0 < p < 1.$$

With the uniform prior, the posterior density is proportional to the likelihood function displayed in Figure 1.1. The difference between classical inference regarding this likelihood function and the Bayesian approach lies in the interpretation of this curve. For a classical statistician, this curve represents the probability of the sample result (5 agreements in a sample size of 16) for different fixed values of the unknown parameter p. For Bayesians, this curve is a representation of beliefs (using the language of probability) about the value of the population proportion.

1.1.9   Posterior densities with alternative priors

One attractive feature of Bayesian inference is that varying prior beliefs about parameter values can be introduced through the prior density. Given any prior, Bayes' theorem provides an automatic mechanism for revising one's posterior probability assessments after obtaining additional evidence in the form of data. To illustrate this point, we next consider two distinct prior distributions in modeling the student survey data. The first prior is chosen because it is convenient analytically, having the same functional form as the likelihood function. Such priors are called conjugate priors and often lead to posterior densities that have a well-studied functional form. The second prior is nonconjugate and leads to a posterior that has a comparatively complex form. However, the second prior density may provide a more accurate representation of some individual's prior belief about the parameter of interest, and so we examine it also with the purpose of demonstrating a more mathematically challenging nonconjugate Bayesian analysis.

A beta prior

One general approach for constructing a prior distribution is to assume that the prior is a member of a parametric family of densities, and to then choose the parameters of the prior so that it approximately represents prior beliefs. For binomial likelihoods, a convenient class of priors for the proportion p is the beta family, which is the conjugate class of priors for binomial likelihoods. Every beta density can be completely described by two parameters, say a and b. Given a and b, a beta(a, b) prior density is proportional to

$$g(p) \propto p^{a-1}(1 - p)^{b-1}, \qquad 0 < p < 1. \qquad (1.3)$$

How does one choose a and b to match one's prior beliefs? A simple way of specifying these parameters is to simply plot various members of this family of densities for different values of a and b, and to then choose that member which best summarizes one's prior beliefs. Alternatively, because the beta density is conjugate for binomial data, we might, instead, imagine a previous survey that would represent the amount of information we feel we have concerning the proportion p. We view this previous survey as a sample of size $a + b$, consisting of a successes and b failures. In our survey example, we might suppose our prior information was equivalent to a preliminary survey in which we observed one person in agreement with the question, and nine people in disagreement. In this case, we are implicitly stating that we have prior information equivalent to a sample survey of 10 individuals, and that our best guess of the true population proportion is $1/(1 + 9) = .1$. This particular information corresponds to a beta$(2, 10)$ prior (pictured in Figure

1.2, bottom), as can be verified by comparing (1.3) to (1.1). The strength of this prior information can be measured by the sum of parameters $a + b$. If we felt we knew more about the true population proportion, we could keep the guess $a/(a+b)$ constant and increase the imaginary sample size $a + b$; if we knew less, we could decrease the value of $a + b$.

In the case where little prior information is available about the proportion, it is often convenient, as in the earlier example, to assume that p has a uniform density on the interval (0, 1). By looking at the general form of the beta density (1.2), we see that that a uniform density is a special case of the beta density with $a = 1$ and $b = 1$.

Because beta densities are conjugate for binomial data, the posterior distribution that is generated with a beta density has a convenient functional form; in this case another beta density. More specifically, if we observe y successes and $n-y$ failures, and employ a beta(a, b) prior, Bayes' theorem leads to a posterior density for p of the form

$$g(p|\text{data}) \propto [p^y(1 - p)^{n-y}][p^{a-1}(1 - p)^{b-1}]$$
$$= p^{a^*-1}(1 - p)^{b^*-1},$$

where $a^* = a + y$ and $b^* = b + n - y$. In other words, the posterior distribution of p is also a beta density with parameters a^* and b^*.

Returning to the student survey example, suppose we assume, as discussed above, a beta prior with parameters $a = 2$ and $b = 10$. After observing 5 agreements (successes) and 11 disagreements (failures) in the survey responses, it then follows that the posterior density is a beta density with parameters $a^* = 2 + 5$ and $b^* = 10 + 11$. Figure 1.3 illustrates this beta prior $g(p)$, the likelihood $L(p) = f(6|p)$, and the posterior density $g(p|$ data). Note that before observing the survey data, the prior suggests that the value of p is likely to lie in a neighborhood of 0.1, while the data indicates (through the likelihood) that the true proportion is likely to be around $p = 0.3$. The posterior density represents a weighted average of these two sources of information and concentrates its mass near .25.

Shrinkage

One way of understanding the combination of information from the prior density and the data is through the notion of shrinkage. The mean of a beta density with parameters a and b is

$$\frac{a}{a + b}. \tag{1.4}$$

After observing s successes and f failures, the posterior mean, denoted by μ_p, is

$$\mu_p = \frac{a + s}{a + b + s + f},$$

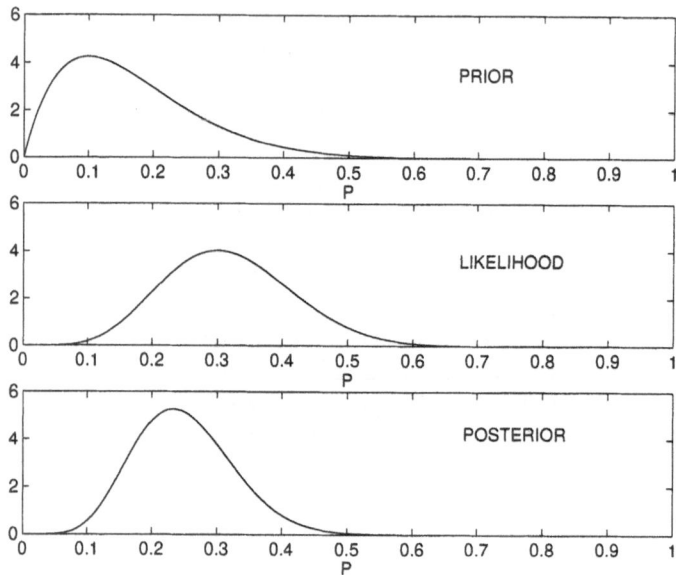

FIGURE 1.3. Prior, likelihood, and posterior using a beta(2, 10) prior and a sample of 5 successes and 11 failures.

which can be reexpressed as

$$\mu_p = \lambda \frac{a}{a+b} + (1-\lambda)\hat{p},$$

where $\frac{a}{a+b}$ is the mean of the prior density, $\hat{p}$ is the proportion of successes in the sample, and $\lambda = \frac{a+b}{a+b+s+f}$ is a fraction between 0 and 1. The posterior mean can be called a shrinkage estimate; it shrinks the observed proportion of successes $\hat{p} = \frac{s}{s+f}$ toward the prior mean $\frac{a}{a+b}$. The degree of shrinkage is controlled by the fraction λ. The value of this fraction depends on the size of the sum of beta prior parameters $a + b$ relative to the sample size $n = s + f$.

A histogram prior

Although the beta prior density is easy to assess and is attractive computationally, it may not provide a good representation of one's prior opinion. An alternative approach to assessing a prior is to construct a cruder form of a probability distribution which better represents prior information about the parameter. For example, we might specify the probabilities that the survey proportion fell into predefined divisions of the unit interval. More specifically, we could divide the interval (0, 1) into the five subintervals (0, .2), (.2, .4), (.4, .6), (.6, .8), and (.8, 1). We could then assign probabilities to each interval in accordance with our prior belief that the population proportion fell into each region. For concreteness, suppose the prior probabilities for these subintervals are 0.5, 0.3, 0.1, .05, and .05, respectively.

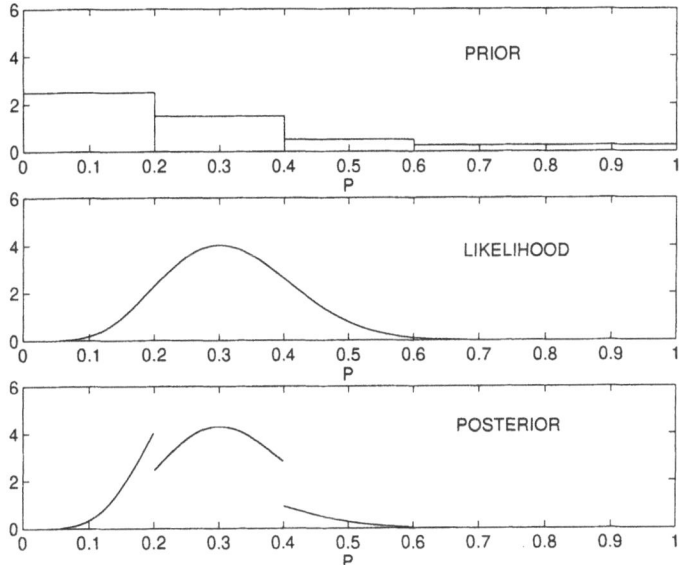

FIGURE 1.4. Prior, likelihood, and posterior using a histogram prior.

The use of this "histogram prior" is illustrated in Figure 1.4 for our example with $s = 5$ and $f = 11$. The top plot depicts the prior, the middle plot the likelihood, and the bottom plot the posterior density. Note that the posterior density has an unusual shape due to the noticeable discontinuities of the prior density at the values $p = .2$ and $p = .4$.

The histogram prior is attractive because it is simple to specify, requires no parametric assumptions, and provides great flexibility in specifying prior beliefs. A disadvantage of this prior is that the resulting posterior density does not have a convenient functional form, and discontinuities in the posterior may not represent actual beliefs. Nonetheless, this example illustrates that the posterior distribution can be computed, via Bayes' rule, for any valid prior and any well-defined likelihood.

1.1.10   Summarizing the posterior density

From the Bayesian perspective, all knowledge about a parameter, after observing data, is contained in the posterior distribution. Thus, a plot of the posterior, as shown in Figures 1.3 and 1.4, is often an effective way of communicating our knowledge about the distribution of the parameter. However, it can be hard for practitioners to understand the main features of a probability density by viewing its graph— particularly for high dimensional parameters—and so methods for summarizing the posterior are desirable. We now examine several measures which summarize key aspects of the posterior density.

Point estimates

Suppose we are interested in a "best guess" of a parameter. In classical statistical inference, this is called a point estimate of the parameter. From a Bayesian view-point, this best guess is usually some "central" value of the posterior density. Three such values are commonly computed—the mean, the mode, and the median.

For a binomial proportion, the *posterior mean* of p is defined by the integral

$$\mu_p = \int_0^1 p\, g(p|\text{data})dp.$$

Similarly, the *posterior mode* is the value of p, here denoted by p_m, at which the posterior density is maximized:

$$g(p_m|\text{data}) = \max_p g(p|\text{data}).$$

This summary value has a close relationship to the MLE. In particular, if a uniform prior density is assumed, the MLE is equal to the posterior mode. The posterior mode is often referred to as the maximum a posteriori estimate, or MAP estimate for short.

A final measure of central tendency is the *posterior median*. For the survey proportion, this is the value of p, say $p_{.5}$, that satisfies

$$\Pr(p < p_{.5}) = .5.$$

To illustrate the computation of these point estimates, suppose we again take a uniform prior (i.e., a beta$(1,1)$ prior) on p and observe $y = 5$, and $n - y = 11$. The posterior density for p is then a beta$(6, 12)$ density proportional to $p^5(1 - p)^{11}$. It follows that the posterior mode is $p_m = 5/16 = .312$ and is equal to the MLE derived above. The posterior mean and median of this distribution are given by $\mu_p = .333$ and $p_{.5} = .327$, respectively. Computation of the posterior median requires evaluation of the integral of a beta density, while the posterior mean can be computed according to (1.4). Note that the mode, mean, and median are approximately equal, indicating that the shape of the posterior density is nearly symmetric.

Interval estimates

It often happens that a single point estimate is not an adequate summary of the information available for a parameter, and in many such cases, a region that contains the parameter with high probability is desired. Such a region can be specified explicitly or can be approximately defined implicitly. In the latter case, a standard deviation is often supplied, with the understanding that the posterior distribution of the parameter (from the Bayesian perspective) or the sampling distribution of the point estimate (from the classical perspective) has an approximate normal distribution. One Bayesian measure of standard error is the *posterior standard deviation* σ_p, defined for a parameter p as

$$\sigma_p = \sqrt{\int_0^1 (p - \mu_p)^2 g(p|\text{data})dp}.$$

A more complete description of the posterior distribution can be provided through a partial listing of the *quantiles* of the posterior density. The qth quantile of the posterior distribution is the value of the parameter p, p_q, such that the probability that p is less than or equal to p_q is equal to q:

$$\Pr(p < p_q) = q.$$

The posterior median is the percentile $p_{.5}$. The quartiles $p_{.25}$ and $p_{.75}$ bound the middle 50% of the posterior distribution. One can effectively summarize the posterior distribution by stating a set of quantiles for a range of values of the left-tail probability q. In the student survey example with the use of a uniform prior, the 5%, 25%, 50%, 75%, and 95% quantiles of the posterior distribution of p are {.166, .255, .327, .405, .522}.

In practice, one is typically interested in an *interval estimate* for the parameter p. We will refer to such an interval as a *probability interval*—an interval which contains the parameter of interest with a specified posterior probability. If the desired probability content is denoted by γ, one such interval can be specified in terms of the posterior quantiles by $(p_{\gamma/2}, p_{1-\gamma/2})$. This particular interval is often called the "equal-tails" interval since the probability that the proportion lies outside the interval on each side is $\gamma/2$. A 90% probability interval for the proportion of students in agreement with the survey statement is $(p_{.05} = .166, p_{.95} = .522)$. The posterior density of the proportion is shown in Figure 1.5, and the limits of this 90% interval, together with the corresponding probability content, are displayed on the plot.

Normal approximations to the posterior density

When the posterior density is approximately bell-shaped, it may be convenient to approximate the posterior density with a normal curve with mean μ_p and standard deviation σ_p. Under this approximation, the probability that the proportion lies within one standard deviation of the mean is .68 and the probability that p lies within two standard deviations of the mean is .95. In general, an approximate 100γ% probability interval for p is given by the interval

$$(\mu_p - z_{\gamma/2}\sigma_p, \mu_p + z_{\gamma/2}\sigma_p),$$

where $z_{\gamma/2}$ is the value of the standard normal distribution with upper-tail area $\gamma/2$.

For the survey example, the posterior mean and standard deviation are given by $\mu_p = .333$ and $\sigma_p = .108$, respectively. From Figure 1.5, we see that the posterior density is approximately Gaussian, and so the normal approximation is likely to be adequate for most purposes. The limits for an approximate 90% probability

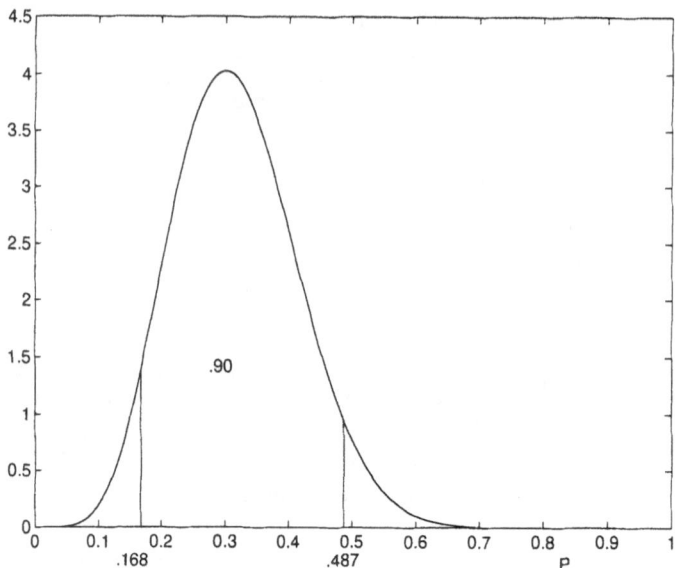

FIGURE 1.5. Posterior density of proportion p with uniform prior and $y = 5, n - y = 11$, and a 90% probability interval indicated by the area between the vertical lines under the density.

interval are thus given by

$$(.333 - 1.645 \times .108, .333 + 1.645 \times .108) = (.155, .511).$$

Note that the endpoints of this interval are close to the exact equal-tails interval quoted above.

Posterior probabilities

In some situations, the computation of posterior probabilities is of interest. For example, in the students survey, interest might focus on the plausibility of the hypothesis that the proportion p of undergraduate Duke students who report that grading policy plays no role in their selection of distribution requirements is greater than .5. Support for this statement can be assessed by computing the posterior probability that $p > .5$. This probability is the area under the posterior density over the interval $(.5, 1)$ or, mathematically,

$$\Pr(p > .5) = \int_{.5}^{1} g(p|\text{data})dp. \tag{1.5}$$

If this posterior probability is small, the data provide evidence against the hypothesis that $p > .5$.

In this case, the posterior density for the proportion of students who reported that grading played no role (assuming a uniform prior) is proportional to $p^5(1 - p)^{11}$. From the incomplete beta distribution (the function which provides values for the integral in (1.5)), we find that $\Pr(p > .5) = .072$. Since this probability is small, we conclude that it is unlikely that a majority of Duke statistics students agree with this statement.

1.1.11 Prediction

We have concentrated on making inferential statements concerning the true value of a parameter. However, in many situations, interest focuses on predicting values of a future sample from the population. In the Duke survey, a professor might conduct an additional survey to assess student opinion concerning the role of grading in enrollment decisions. Interest might then focus on predicting the number of students in the new sample who report that grading plays no role in such decisions.

To pursue this problem further, suppose that an additional survey is conducted and that a random sample of size n^* is to be collected. Let y^* denote the unknown number of students in this sample who agree with the survey statement of "significance," and let $n^* - y^*$ be the number of students who disagree. If we *knew* the value of the population proportion p based on the earlier survey (we do not!), then the probability that we observe y^* students in agreement with the statement would then be given by the same binomial probability as in (1.1):

$$f(y^*|p) = \binom{n^*}{y^*} p^{y^*}(1 - p)^{n^* - y^*}, \qquad y^* = 0, ..., n^*.$$

However, we do not know the value of the proportion p, but, instead, we are able to summarize our beliefs about this parameter by the posterior distribution obtained from the previous survey, $g(p|\text{data})$, where "data" refers to the number of agreements and disagreements in the previous survey. Then, the *predictive probability* of y^* (in a future sample of size n^*) is given by the integral

$$f(y^*|\text{data}) = \int_0^1 f(y^*|p)g(p|\text{data})dp, \qquad y^* = 0, ..., n^*.$$

From our initial survey, assuming a uniform prior density, the posterior distribution was a beta(6,12) density proportional to $p^5(1 - p)^{11}$. To predict the number of students y^* in agreement with the statement in a future sample of size $n^* = 20$, we must calculate the value of the integral

$$f(y^*|\text{data}) = \int_0^1 f(y^*|p)beta(p; 6, 12)\, dp, \qquad y^* = 0, ..., n^*,$$

where $beta(|a, b)$ is the beta density (1.2) with parameters a and b. Figure 1.6 plots this predictive density of y^* for values between 0 and 20. Note that the most probable value is $y^* = 6$ in this sample of size 20.

As in parameter estimation, we can construct a probability set for this unknown future observation. This interval is constructed by first ordering the values of y^* from most probable to least probable. If we desire a 90% prediction interval, we add the most probable values into our probability set until the total probability of the set exceeds .9. In this example, the 90% prediction set is $\{2, 3, ..., 11\}$.

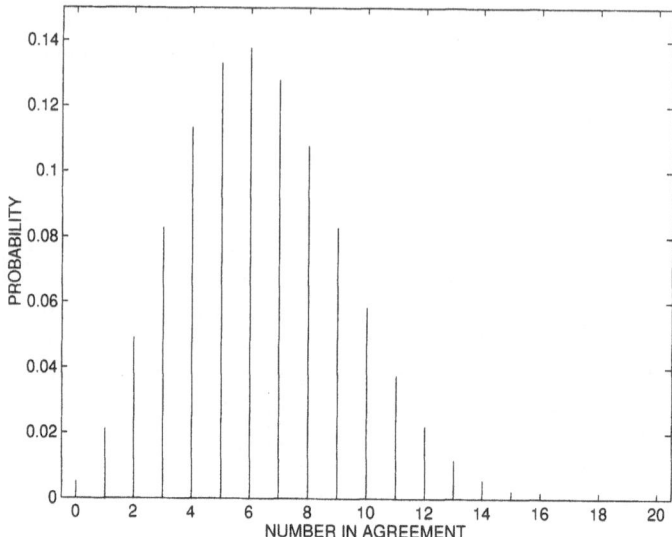

FIGURE 1.6. The predictive density of the number of students in agreement in a future sample of size 20.

This interval is rather broad, reflecting both the binomial variation in the future observation and the posterior uncertainty in the true value of p.

1.2 Inference for a normal mean

To this point, we have confined our attention to the analysis of binomial data and inference on the binomial proportion. We now turn our attention to inference for normally distributed data when both the population mean and variance are unknown.

To motivate our studies, suppose that we are interested in inference regarding a normal mean and that our data are a sample of $n = 13$ measurements of the NOAA/EPA ultraviolet index measured in Los Angeles. The 13 data points were collected from archival data maintained by the Climate Prediction Center of the National Oceanic and Atmospheric Administration (NOAA) and were arbitrarily extracted for every Sunday falling in October during the years 1995–1997. This particular index is assigned non-negative integer values, and different values are associated with different exposure levels. Values in the range 0–2 are considered by the Environmental Protection Agency (EPA) to represent minimal exposure, values of 3–4 represent low exposure, values of 5–6 moderate exposure, values of 7–9 high exposure, and values of 10 and greater very high exposure.

A simple model for this type data can be obtained by assuming that each observation represents an independent realization from a normal distribution with mean

μ and variance σ^2. Under this assumption, the parameter of interest is the average October UV index in Los Angeles, represented by the parameter μ.

1.2.1   A classical analysis

Let the n measurements of the UV index be denoted by $y_1, ..., y_n$; the specific values of the measurements were

$$7, 6, 5, 5, 3, 6, 5, 5, 3, 5, 5, 4, 4.$$

Under a normal, or Gaussian, model for the data, the density function for one observation, given the values of the population mean μ and the population variance σ^2, is

$$f(y_i|\mu, \sigma^2) = \frac{1}{\sqrt{2\pi}\sigma} \exp\left\{-\frac{1}{2\sigma^2}(y_i - \mu)^2\right\}, \qquad -\infty < y_i < \infty.$$

Thus, to construct the joint probability density function for the entire sample of *independent*[3] measurements, we multiply the marginal densities of the individual observations to obtain

$$f(y_1, ..., y_n|\mu, \sigma^2) = \prod_{i=1}^{n} f(y_i|\mu, \sigma^2)$$

$$= \prod_{i=1}^{n} \frac{1}{\sqrt{2\pi}\sigma} \exp\left\{-\frac{1}{2\sigma^2}(y_i - \mu)^2\right\}. \qquad (1.6)$$

Once the measurements $y_1, ..., y_n$ are made, we obtain the likelihood function for the parameters μ and σ^2 by viewing (1.5) as the joint density function of the parameters, for the fixed values of the data. The likelihood function is then given by

$$L(\mu, \sigma^2) = \prod_{i=1}^{n} \frac{1}{\sqrt{2\pi\sigma^2}} \exp\left\{-\frac{1}{2\sigma^2}(y_i - \mu)^2\right\}, \qquad -\infty < \mu < \infty, \quad \sigma^2 > 0.$$

Based on the likelihood function, estimates of the parameters μ and σ^2 can be obtained using the method of maximum likelihood, as discussed above.

In the maximization of the likelihood function, it is convenient (as in the proportion example) to first take the logarithm of the likelihood function. We differentiate

---

[3]The assumption of independence may not be valid for these data, since measurements taken at weekly intervals are likely to be correlated. In other words, if the observation recorded on the first Sunday in October in a given year is above (below) the mean μ, the observation recorded on the second Sunday $k + 1$ is also likely to be above (below) the mean. This effect is likely to persist even when a trend is fit to account for the fact that the UV index tends to decrease late in October, due to the onset of winter. For present purposes, we can ignore the later trend, since we are interested only in the mean October UV index value, and for simplicity, we do ignore the small correlation that probably exists between consecutive Sundays.

the log-likelihood $\log L(\mu, \sigma^2)$ with respect to both μ and σ^2, and determine the values of the parameters that make the partial derivatives equal to 0. Doing this leads to the following equations that implicitly determine the MLE:

$$\frac{1}{\sigma^2} \sum_{i=1}^{n} (y_i - \mu) = 0$$

$$-\frac{n}{\sigma^2} + \frac{1}{\sigma^4} \sum_{i=1}^{n} (y_i - \mu)^2 = 0$$

The solutions to these equations are $\hat{\mu} = \bar{y}$ and $\hat{\sigma}^2 = \sum_{i=1}^{n} (y_i - \bar{y})^2/n$, where $\bar{y} = \sum_{i=1}^{n} y_i/n$ is the sample mean.

As in our previous example of estimating a binomial proportion, we could base inference on a normal approximation to the sampling distribution of the MLE. However, in this model, the exact sampling distributions for both the estimates $\hat{\mu}$ and $\hat{\sigma}^2$ are known. Furthermore, the sampling distribution of the statistic

$$T = \frac{\bar{y} - \mu}{s/\sqrt{n}}$$

has a t distribution with $n - 1$ degrees of freedom, where

$$s = \sqrt{\sum_{i=1}^{n} (y_i - \bar{y})^2/(n - 1)}$$

is the sample standard deviation. This sampling distribution is used in the construction of both confidence intervals and hypothesis tests for the mean μ. For example, by inverting the probability statement

$$\Pr\left(-t_{(1-\gamma)/2} < \frac{\bar{y} - \mu}{s/\sqrt{n}} < t_{(1-\gamma)/2}\right) = \gamma,$$

where $t_{(1-\gamma)/2}$ is the $(1 - \gamma)/2$ quantile of a $t(n - 1)$ curve, we obtain the interval estimate

$$(\bar{y} - t_{(1-\gamma)/2}\, s/\sqrt{n}, \bar{y} + t_{(1-\gamma)/2}\, s/\sqrt{n}).$$

This interval is called a 100γ confidence interval for μ; the probability that the random interval contains the unknown population mean μ in repeated sampling is equal to γ.

The sampling distribution of the statistic

$$W = \frac{(n - 1)s^2}{\sigma^2}$$

is a χ^2 random variable with $n - 1$ degrees of freedom. Like the statistic T, the statistic W may be used to either obtain confidence intervals for the variance σ^2 or to conduct statistical tests of hypotheses concerning the "true" value of σ^2.

1.2.2   Bayesian analysis

The Bayesian analysis of the normal mean problem requires the specification of a joint prior density for μ and σ^2. Although any informative prior distribution might be used, for present purposes we consider only the standard noninformative, or "reference," prior density, which is proportional to

$$g(\mu, \sigma^2) = \frac{1}{\sigma^2}, \qquad -\infty < \mu < \infty, \qquad \sigma^2 > 0.$$

Having specified the prior, the posterior density is determined (up to a proportionality constant) by multiplying the prior density by the likelihood function. Using the likelihood function specified in (1.6), one obtains the following posterior density of the mean and variance:

$$g(\mu, \sigma^2 | \text{data}) \propto g(\mu, \sigma^2) \, f(y_1, \, ..., \, y_n | \mu, \sigma^2)$$

$$= \frac{1}{\sigma^2} \prod_{i=1}^{n} \frac{1}{\sqrt{2\pi\sigma^2}} \exp\left\{-\frac{1}{2\sigma^2}(y_i - \mu)^2\right\}.$$

In the above expression, "data" refers to the observations $\{y_1, \, ..., \, y_n\}$. Simplifying this expression yields

$$g(\mu, \sigma^2 | \text{data}) \propto \frac{1}{(\sigma^2)^{n/2+1}} \exp\left\{-\frac{1}{2\sigma^2}(S + n(\mu - \bar{y})^2)\right\}$$

$$= \left[\frac{1}{(\sigma^2)^{(n-1)/2+1}} \exp\left\{-\frac{S}{2\sigma^2}\right\}\right]$$

$$\times \left[\frac{1}{\sigma} \exp\left\{-\frac{1}{2\sigma^2}n(\mu - \bar{y})^2\right\}\right],$$

where $\bar{y}$ is the mean of the observations and $S = \sum_{i-1}^{n}(y_i - \bar{y})^2$ is the sum of the squared deviations of the observations about the mean.

The functional form of the joint posterior can be identified from the last expression above. Suppose that the variance σ^2 is fixed. Then, the posterior density of the mean μ, conditional on this variance, is proportional to

$$g(\mu | \sigma^2, \text{data}) \propto \frac{1}{\sigma} \exp\left\{-\frac{1}{2\sigma^2}n(\mu - \bar{y})^2\right\},$$

which is recognizable as a normal density with mean $\bar{y}$ and standard deviation $\sigma/\sqrt{n}$. The first term in brackets,

$$g(\sigma^2 | \text{data}) \propto \frac{1}{(\sigma^2)^{(n-1)/2+1}} \exp\left\{-\frac{S}{2\sigma^2}\right\},$$

is the marginal posterior density of the variance σ^2. This is the distribution of an inverse gamma distribution with shape parameter $(n - 1)/2$ and scale parameter $S/2$. The joint distribution for (μ, σ^2) is called a normal/inverse-gamma form.

Recall that we were interested in inference about the mean parameter μ. From a classical perspective, it can be difficult to make inferences about one parameter, such as μ, when a second parameter, such as the variance σ^2, is unknown. The second parameter not of direct interest is called a nuisance parameter, since there is no standard classical method for removing this parameter from the likelihood function. However, there is a standard method of removing this nuisance param-eter in a Bayesian analysis. Given the joint posterior distribution on (μ, σ^2), one bases Bayesian inferences about the mean parameter μ on its marginal posterior density $g(\mu|\text{data})$. By using rules of probability, this marginal density is obtained by integrating over the variance parameter σ^2 in the joint posterior density:

$$g(\mu|\text{data}) = \int g(\mu, \sigma^2|\text{data}) \, d\sigma^2.$$

In the normal error model, the integration over the variance parameter can be performed analytically. The result is a scaled t distribution. By rescaling, we find that the posterior distribution of the standardized quantity

$$\frac{\mu - \bar{y}}{\sqrt{S/(n(n-1))}} = \frac{\mu - \bar{y}}{s/\sqrt{n}}$$

has a t distribution on $n - 1$ degrees of freedom.

This distributional result for the posterior distribution of the normal mean μ with a noninformative prior resembles the result for the sampling distribution of the t-statistic discussed in the previous section. This means that the 95% interval estimate for the mean μ given by

$$(\bar{y} - t\, s/\sqrt{n}, \bar{y} + t\, s/\sqrt{n})$$

has a dual interpretation. The probability that this interval covers the fixed, but unknown, mean μ in repeated sampling is .95. However, it also may be interpreted from a Bayesian perspective as the probability that the true mean μ falls in the observed interval is .95.

Returning to the UV index data, the sample mean of the 13 measurements was $\bar{y} = 4.85$, and the standard error was se $= s/\sqrt{n} = 0.32$. In this case, μ represents the mean October UV index value in Los Angeles, and the posterior density of

$$\frac{\mu - 4.85}{0.32}$$

has a t distribution with $n - 1 = 12$ degrees of freedom. The posterior density of μ is displayed in Figure 1.7.

As in the survey example, the posterior density can be summarized in a variety of ways, depending on the goals of the analysis. Because the posterior density is symmetric, the obvious point estimate of the population mean is $\bar{y} = 4.85$, which corresponds to the posterior mean, mode, and median. To construct a 95% probability interval for μ, we use a t distribution on 12 degrees of freedom, yielding

$$(4.85 - t\,(0.32), 4.85 + t\,(0.32)) = (4.16, 5.54),$$

where t is the value of a t random variable with 12 degrees of freedom with upper-tail probability .025.

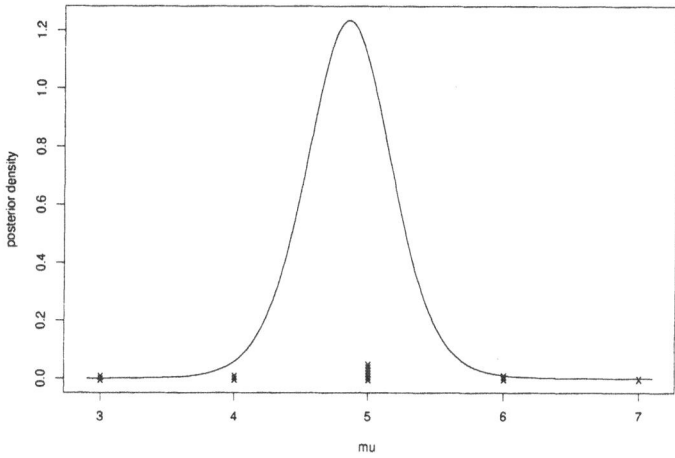

FIGURE 1.7. Marginal posterior density of the mean October UV index in Los Angeles. The observations are indicated by the solid dots.

Direct computation of posterior probability intervals might also be used to summarize the ultraviolet exposure in Los Angeles during October. For example, exposure is considered "high" when the UV index exceeds 6.5. We can directly assess the probability that the mean noontime UV exposure is high by computing the posterior probability that μ exceeds 6.5, which, in this case, can be computed from tabled values of t distributions as

$$Pr(\mu > 6.5) = 0.00011.$$

Thus, there is only a 1/10,000 chance that the mean UV index in Los Angeles in October is "high" by EPA standards.

1.3   Inference about a set of proportions

The normal mean problem illustrates the application of Bayesian inference in the situation where there are two unknown parameters, the mean and variance, and one is primarily interested in learning about the mean. In this section, we illustrate a Bayesian approach for simultaneously estimating a set of binomial proportions. The prior distribution that will be used in this application reflects the initial belief that the proportions are similar in size.

Tsutakawa et al. (1985) discuss the problem of simultaneously estimating the rates of death from stomach cancer for males at risk in the age bracket 45–64 for the largest cities in Missouri. Table 1.2 displays the mortality data for 20 of these cities. The number n_i at risk and the number of cancer deaths y_i in city i are displayed in the table. We assume that $\{y_i\}$ are independent binomial variables with sample sizes $\{n_i\}$ and probabilities of death $\{p_i\}$. The probability p_i represents the

(0, 1083)	(0, 855)	(2, 3461)	(0, 657)	(1, 1208)	(1, 1025)
(0, 527)	(2, 1668)	(1, 583)	(3, 582)	(0, 917)	(1, 857)
(1, 680)	(1, 917)	(54, 53637)	(0, 874)	(0, 395)	(1, 581)
(3, 588)	(0, 383)				

TABLE 1.2. Cancer mortality data. Each ordered pair represents the number of cancer deaths y_i and the number at risk n_i for an individual city in Missouri.

hypothetical proportion of an infinite population of men at risk in the ith city that would die from stomach cancer.

The objective of this study is to estimate the probabilities of death for all 20 cities. The standard method of estimating the proportions $p_1, ..., p_{20}$ is based on the method of maximum likelihood. The likelihood function in this case is the probability of observing the data $\{(y_i, n_i)\}$ viewed as a function of the proportions. Since the $\{y_i\}$ are independent binomial variables, the probability of these data is given by the product of binomial densities, and so the likelihood function is given by

$$L(p_1, ..., p_{20}) = \prod_{i=1}^{20} \binom{n_i}{y_i} p_i^{y_i}(1 - p_i)^{n_i - y_i}.$$

The maximum likelihood estimates of $p_1, ... p_{20}$ are the sample proportions

$$\hat{p}_1 = \frac{y_1}{n_1}, ..., \hat{p}_{20} = \frac{y_{20}}{n_{20}}.$$

These classical estimates of the population proportions are troublesome for cities in which there are no cancer deaths. For these cities, $y_i = 0$ and the estimated probability of death from stomach cancer is $\hat{p}_i = 0/n_i = 0$. Clearly, this is an unreasonably optimistic estimate of the population proportion. In such cases, a more reasonable estimate of the probability of cancer death would be a small nonzero value which is similar to the estimated probability of death for the other cities. In other words, it makes sense to estimate the probability of cancer death for a particular city by a value which combines the information from that city with the information collected from other cities.

To model the fact that the probability of death from stomach cancer is similar from one city to the next, it is necessary to construct a multivariate prior density for the set of proportions $\{p_1, ..., p_{20}\}$. One way to construct such a prior distribution is to assume that $p_1, ..., p_{20}$ are drawn independently from beta(a_i, b_i) densities, where the prior parameters a_i and b_i are chosen to reflect prior beliefs about the value of the ith population proportion. This process can be difficult since it requires the assessment of $2 \times 20 = 40$ parameters, 2 for each proportion.

An attractive alternative to assigning a prior density to each of these proportions individually is premised on the belief that the 20 proportions have similar values, or that they were at least drawn a priori from a common distribution. In this example, one might posit that stomach cancer among these at-risk individuals is a rare event for all 20 cities in the study and, furthermore, that the probabilities of cancer are approximately the same for all cities in Missouri. If this is the case, we say that our prior knowledge concerning the proportions is *exchangeable*. From a practical standpoint, exchangeability implies that one believes that the sizes of the 20 proportions are similar.

An exchangeable prior distribution for the proportions can be constructed in two stages. In stage one, we assume a common prior density for all city proportions. The parameters of this first-stage prior are left unspecified. In the second stage, we assign a prior distribution on the unknown hyperparameters in the first stage.

In our example, this two-stage process can be implemented as follows. First, we model the belief that the 20 cancer probabilities have similar values by assuming (stage one) that $p_1, ..., p_{20}$ are independently distributed according to a common beta distribution with parameters a and b. It is helpful to reparameterize the hyperparameters a and b to new values which reflect the location and spread of the beta prior. The prior mean m and prior precision parameter K are defined by

$$m = \frac{a}{a + b}, \qquad K = a + b.$$

The hyperparameter m represents a guess at the location of each proportion and K represents the information content of this guess, measured in terms of prior observations.

Since the beta parameters m and K are unknown, we use the second stage of the model to reflect our beliefs about the locations of the common mean and precision for the 20 proportions. In practice, the distributions for these two parameters will be chosen to be relatively diffuse, reflecting modest prior information. In this example, we assume m and K are independent, with

$$g(m) \propto m^{.01-1}(1 - m)^{9.9-1}, \quad 0 < m < 1, \quad g(K) = \frac{1}{(1 + K)^2}, \quad K > 0.$$

Thus, the parameter m is assigned a beta density with mean .001 and precision 10. This means that we believe all of the proportions are close to .001, but this belief is only worth about 10 observations. (The value .001 is the combined proportion of deaths for all cities.) The precision parameter K is assigned the vague density $1/(1 + K)^2$, reflecting little knowledge about the location of this parameter.

This two-stage construction process leads to a multivariate prior distribution for $p_1, ..., p_{20}$ which reflects the belief that the sizes of the proportions are similar. Figure 1.8 plots a sample of 1000 "representative" values from the prior distribution of 2 of the proportions. Since a display of the proportion values is not helpful because most of the values are close to 0, this figure plots the distribution of the logit of $p_1 = \log \frac{p_1}{1-p_1}$ against the logit of $p_2 = \log \frac{p_2}{1-p_2}$. From this plot, we note that the two proportions are positively correlated—this correlation is a consequence of the belief that the two proportions are similar in size.

The computation of the posterior distribution of the proportions $p_1, ..., p_{20}$ is more difficult here than in the one-proportion situation. Section 2.7 illustrates an algorithm for summarizing the multivariate posterior distribution by use of simulation methodology. In this chapter, we simply describe the behavior of the

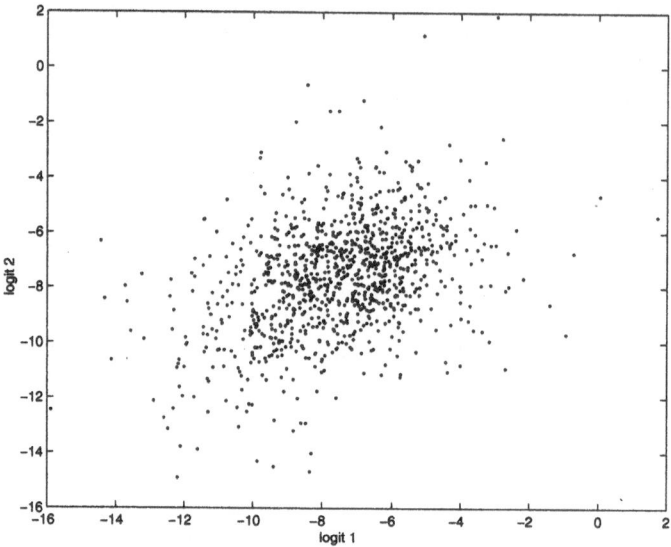

FIGURE 1.8. A sample of 1000 representative values from the joint prior distribution of the logit of p_1 against the logit of p_2.

posterior distribution for the cancer mortality data assuming that the posterior distribution has already been effectively summarized.

Table 1.3 displays the observed number of cancer deaths y_i, the number at risk n_i, and the observed proportion $\hat{p}_i$ for all 20 cities. The values of $\hat{p}_i$ are estimates under a model which does not assume any relationship between the proportions. These estimates are widely dispersed in their relative values (ranging between 0 and .0051), indicating the effect of the small numbers of observed deaths in the sample. As an alternative to assuming that all proportions were independent, we might, instead, assume that the probabilities of death across cities were identical ($p_1 = \cdots = p_{20}$). In this model, we would estimate the common probability of death by the pooled estimate

$$\bar{p} = \frac{y_1 + \cdots + y_{20}}{n_1 + \cdots + n_{20}} = .000993.$$

The estimates under the exchangeability model present a compromise between the estimates obtained by assuming that the probabilities of death from stomach cancer were completely unrelated to one another in different cities, and estimates obtained under the model in which the proportions are assumed to be identical across all cities. Bayesian estimates obtained from the two-stage exchangeability model are depicted in the last column of Table 1.3. These point estimates represent the posterior means of the probabilities under the model specified above. The posterior means "shrink" or adjust the observed proportions $\hat{p}_i$ toward the pooled estimate $\bar{p}$. For example, the posterior mean of the cancer probability for the first city is .0007, which lies between the observed proportion $\hat{p}_1 = 0$ and the

y_i	n_i	$\hat{p}_i$	Posterior mean
0	1083	0	.0007
0	855	0	.0008
2	3461	.0006	.0008
0	657	0	.0008
1	1208	.0008	.0010
1	1025	.0010	.0011
0	527	0	.0009
2	1668	.0012	.0011
1	583	.0017	.0012
3	582	.0052	.0020
0	917	0	.0008
1	857	.0012	.0011
1	680	.0015	.0012
1	917	.0011	.0011
54	53637	.0010	.0010
0	874	0	.0008
0	395	0	.0009
1	581	.0017	.0012
3	588	.0051	.0020
0	383	0	.0009

TABLE 1.3. Observed number of deaths y_i, number at risk n_i, observed proportion $\hat{y}_i$, and Bayes posterior mean using exchangeable prior for cancer mortality example.

pooled estimate $\bar{p} = .000993$. All the posterior means exhibit this same shrinking behavior, and the amount of shrinkage depends on the sample size for the particular city. For example, the observed proportion for city 2, with a sample size of $n_2 = 855$, is shrunk more toward the pooled estimate than the observed proportion for city 1, which had a sample size of $n_1 = 1083$. In general, incorporating information about model parameters in this way provides a mechanism for combining related, but not identical, experiments.

1.4 Further reading

Readers seeking elementary statistics texts written from a Bayesian viewpoint can refer to *Statistics: A Bayesian Perspective* by Donald Berry or at a somewhat more advanced level, *Bayesian Statistics* by Peter Lee. Intermediate-level references include *Bayesian Inference in Statistical Analysis* by Box and Tiao and *Bayesian Data Analysis* by Gelman, Carlin, Stern, and Rubin. At an advanced graduate level, both Christian Robert's *The Bayesian Choice* and Bernardo and Smith's *Bayesian Theory* are recommended.

1.5   Exercises

1. In 1986, the St. Louis Post Dispatch was interested in measuring public support for the construction of an indoor stadium. The newspaper conducted a survey in which they interviewed 301 registered voters. In this sample, 135 voters opposed the construction of a new stadium. Let p denote the proportion of all registered voters in the St. Louis voting district opposed to the stadium.

   a. Given that 135 out of 301 in the survey opposed the construction of a new stadium, compute the likelihood function $L(p)$ for values of $p = .1, .2, .3, .4$, and $.5$.

   b. Compute the maximum likelihood estimate $\hat{p}$ and the associated standard error $se(\hat{p})$.

   c. Using your answers from (a) and (b), find an approximate 90% confidence interval for p.

2. (Continuation of Exercise 1) Consider estimation of p from a Bayesian viewpoint.

   a. Suppose that you place a uniform prior on the unknown proportion. Find the posterior density of p and graph the density for values of the proportion between 0 and 1.

   b. Show that the posterior density is a beta density.

   c. Find the posterior mean of p. Compare your estimate with the MLE computed in Exercise 1.

   d. Find a 90% probability interval for p. Compare your interval with the classical 90% confidence interval computed in Exercise 1.

3. (Continuation of Exercise 1) Suppose that a newspaper reporter has some prior knowledge about the support of the indoor stadium from a small survey taken the previous month. She represents her opinion about the proportion p by means of the informative prior density

$$g(p) \propto p^5(1 - p)^5, \qquad 0 < p < 1.$$

   a. Graph this density function. What does this density say about the opinion of the newspaper reporter regarding the proportion of voters in favor of the indoor stadium?

   b. Find the posterior density of p.

   c. Compute the posterior mean of p.

   d. Find a 90% probability interval for p.

   e. Compare the Bayesian point estimate and interval estimate found in (c) and (d) with the estimates found in Exercise 2 using a uniform prior distribution. Discuss the influence of the informative prior density on the posterior inference in this case.

4. (Continuation of Exercise 1) Suppose a city councilman wants to increase the city sales tax to pay for the construction of the new stadium. The councilman

claims that under half of the registered voters are opposed to the stadium construction. She would like to use the sample survey of the newspaper to test the two hypotheses $H : p \geq .5$, $K : p < .5$, where p is the proportion of all registered voters in St. Louis opposed to the stadium.

a. A classical method of testing these hypotheses is based on a p-value. In the sample, 301 voters were questioned and 135 were opposed to the stadium. The p-value is the probability of observing this sample result or more extreme if, indeed, exactly half of the registered voters in St. Louis were opposed to the construction; that is,

$$p - \text{value} = \Pr(y \leq 135 | p = .5),$$

where y is a binomial random variable with sample size $n = 135$ and probability of success $p = .5$. Compute the p-value for this example. If this probability is small, say under 5%, then one concludes that there is significant evidence in support of the hypothesis $K : p < .5$.

b. Now consider a Bayesian approach to testing these hypotheses. Suppose, as in Exercise 2, that p is assigned a uniform prior. The posterior density has a beta form and one compares H and K by computing the posterior probability of each hypothesis. We decide in support of the hypothesis K if its posterior probability is sufficiently large or, equivalently, if the posterior probability of the hypothesis H is sufficiently small.
Compute the posterior probability of H

$$\Pr(H|y) = \Pr(p \geq .5|y).$$

Show that this posterior probability is approximately equal to the p-value computed in part (a). This shows that (in the one-sided testing situation) a classical p-value can be given a Bayesian interpretation.

5. (Continuation of Exercise 1) Suppose, as in Exercise 2, that a uniform prior is placed on the proportion p, and that from a random sample of 301 voters, 135 oppose the construction of a new stadium. Also, suppose that the newspaper plans on taking a new survey of m voters. Let y^* denote the number in this new sample who oppose construction.

a. Find the formula for the predictive probability of y^* in a sample of size m.
[Hint: To compute the predictive probability, one must compute the integral of a beta density. In general, the integral of the kernal of a beta density, called the beta function, is given by

$$B(a, b) = \int_0^1 p^{a-1}(1 - p)^{b-1} dp,$$

and so the normalized expression for a beta density is $g(p) = \frac{1}{B(a,b)} p^{a-1}(1 - p)^{b-1}$. The predictive probability of interest can be expressed as a ratio of beta functions.]

b. Suppose 20 voters are to be sampled, so $m = 20$. Find the probability that at least half of this sample will oppose construction. (Use the MATLAB program **p_beta_p** that is supplied in the teaching Bayes toolbox.)

c. Find a 90% prediction interval for y^*.

d. Use the answer to part (c) to find a 90% prediction interval for the *proportion* of the future sample that will oppose construction.

6. Suppose that a company wishes to evaluate the popularity of two razors that they manufacture, which will be called razor A and razor B. A sample of 100 men is asked to shave one side of their face with razor A and the other side with razor B. Let p denote the proportion of all adult men who prefer razor A. The company will discontinue the production of one of the razors if it is significantly less popular than the other. Rather than regard the proportion of men favoring razor A to razor B as a continuous parameter, they simplify their decision problem by assuming that the true value of p can take on one of three values: (1) $p = .4$ (razor B is significantly better), (2) $p = .5$ (the two razors are equally popular), and (3) $p = .6$ (razor A is significantly better). Suppose that 35 men out of 100 prefer razor A.

a. Suppose that the company believes that the three alternatives stated above are equally likely. Thus, the prior distribution on the proportion p assigns equal probability to the values $p = .4, .5,$ and $.6$. Find the posterior distribution on p. What is the updated probability that the two razors are equally popular?

b. Suppose now that the company knows that there are very small differences in the style and quality of the two razors and therefore assigns a prior probability of .8 on the value $p = .5$, while the remaining prior probability is split equally between the values of $p = .4$ and $.6$. Find the posterior distribution of p under this prior assumption and compare this distribution with the posterior distribution based on a uniform prior in part (a).

7. (From Antleman, 1997). Suppose that a trucking company owns a large fleet of well-maintained trucks and assume that breakdowns appear to occur at random times. The president of the company is interested in learning about the daily rate R at which breakdowns occur. (Realistically, each truck would have a breakdown rate that depends possibly on its type, age, condition, driver, and usage. The breakdown rate of the company can be viewed as the sum of the breakdown rate of the individual trucks.) For a give value of the rate parameter R, it is known that the number of breakdowns y on a particular day has a Poisson distribution with mean R:

$$f(y|R) = \frac{e^{-R} R^y}{y!}, \qquad y = 0, 1, 2, \dots .$$

a. Suppose that one observes the number of truck breakdowns for n consecutive days—denote these numbers by $y_1, \dots, y_n$. If one assumes that these are independent measurements, find the joint probability distribution of $y_1, \dots, y_n$.

b. The number of breakdowns for 5 days are recorded to be 2, 5, 1, 0, and 3. Find the likelihood function $L(R)$ of the rate parameter R for these observations. Graph this function by computing the likelihood for the values $R = .1, .5, 1, 2, 4, 8,$ and16 and connecting the points with a smooth curve.

c. Find the MLE of R and its associated asymptotic standard error.

d. Use the results in part (c) to find an approximate 95% confidence interval for R.

8. (Continuation of Exercise 7) Suppose that an employee of the trucking firm wishes to learn about the breakdown rate. She is uncertain about the value of R and so assigns the noninformative prior density

$$p(R) = \frac{1}{R}, \qquad R > 0.$$

a. Find and graph the posterior density of R. Show that it is a member of the gamma family.

b. Find the posterior mean of R.

c. Find a 95% probability interval for R.

d. Compare the Bayesian point and interval estimates with the classical estimates found in Exercise 7.

9. (Continuation of Exercise 7) The president has some knowledge about the location of the Poisson rate parameter R based on the observed number of breakdowns from previous years. His prior beliefs about R are represented by means of the gamma density

$$g(R) \propto R^{4-1} \exp\{-2R\}, \qquad R > 0.$$

a. Plot this prior density. Based on this plot, describe the president's prior beliefs about the rate parameter R.

b. Find and graph the posterior density of R. Show that this density is a special case of a gamma density. [Note: This exercise illustrates the fact that the gamma family of densities is a conjugate class of prior distributions for learning about a Poisson mean.]

c. Find the posterior mean and 95% probability interval for R. Contrast these estimates with the estimates found in Exercise 8 using a noninformative prior.

10. Consider the following example of acceptance sampling. Suppose a company receives a shipment of components from a vendor. The shipment is made in lots of 20 components and the company is unsure of the number D of "defective" components in a particular lot. The number of defectives could conceivably be any number from 0 to 20. To learn about the number D, the company will take a random sample of five components from the lot without replacement, inspect each component in the sample, and record x the number of defectives.

a. Given that the lot contains D defectives, find the probability of observing x defectives in the sample of five.

b. Suppose that $x = 2$ defectives are found in the sample of five. Find the likelihood function $L(D)$.

c. Find the MLE of D.

d. Consider this inference problem from a Bayesian approach. Suppose that the number of defective components D is assigned a uniform prior on the set $\{0, 1, ..., 20\}$. If $x = 2$ defectives are observed, find the posterior density of D.

e. Find the posterior mean and mode of D. Compare these Bayesian estimates with the MLE found in part (c).

f. Suppose that the lot of parts is considered acceptable if the probability of a "small" number of defectives $\Pr(D \le 5)$ is .9 or above. Find the posterior probability of $\{D \le 5\}$ and decide if this particular lot is acceptable.

11. At the end of Section 1.2, we found that the probability that the mean October UV index in Los Angeles was greater than 6.5 was about 0.0001. Using the model of that section, what is the probability that the UV index for a randomly selected day in October exceeds 6.5?

12. Returning to the student survey, 47 students indicated on the survey that the course they were scheduled to take next was taken to satisfy an elective. Of these 47 students, 26 reported that the expected grade distribution in the elective had not influenced their decision to enroll.

a. Choose an appropriate prior for the proportion q of students taking electives at Duke who report that grade policy does not influence enrollment decision. Using this prior, determine the posterior distribution on q and summarize it either analytically or graphically.

b. Compare your posterior distribution on q to the posterior distribution on p, the proportion of students whose enrollment decision in distributional requirements was unaffected by grade distributions, obtained using a uniform prior. Plot both densities in the same plot, and compare their posterior means and modes.

c. Suppose that you were interested in the posterior distribution on the quantity $q - p$, again using the priors in parts (a) and (b). How would you go about determining this posterior distribution? Can you find an analytical expression for it? If not, how might you approximate it?

2

Review of Bayesian Computation

In Chapter 1, we explored the univariate posterior density of a binomial probability and the joint posterior density for the mean and variance of a normal distribution. When conjugate priors were specified in these models, the posterior densities had a form that could be easily summarized. Unfortunately, these simple models are not typical of real-world applications in which statistical models are necessarily more complex and in which posterior densities seldom have analytically tractable forms. Because these more complicated density functions cannot be treated analytically, alternative methods must be invoked.

In this chapter, we address the problem of summarizing complicated and often high-dimensional posterior densities when analytical methods are unavailable. Generally speaking, this problem inevitably reduces to numerically integrating a function that is proportional to the joint posterior density.

Prior to about 1990, numerical integration of posterior densities relied on standard methods of numerical analysis. Two such methods are discussed here. The first exploits the fact that many posterior densities are shaped like multivariate Gaussian densities. The second method involves forming weighted averages of points evaluated on a grid. Grid-based methods can be made arbitrarily precise but, due to computational constraints, are effective only in problems involving low-dimensional (10 or smaller) parameter vectors.

The remainder of the chapter illustrates techniques for summarizing posterior distributions through the use of modern simulation algorithms. To introduce this topic, we begin with the ideal situation in which direct simulation is possible and demonstrate how simulated samples can be used for purposes of statistical inference. We then discuss the more realistic situation in which direct simulation from the posterior density is not possible and introduce Markov chain Monte Carlo (MCMC) methods.

2.1   Integrals, integrals, integrals, . . .

A common theme encountered when implementing the Bayesian paradigm is the necessity to evaluate high-dimensional integrals. Typically, the integrand represents a product of the joint posterior distribution on, say, a k-dimensional parameter vector $\theta = \{\theta_1, \theta_2, \ldots, \theta_k\}$ and some function of θ. Let $\mathbf{y}$ denote the vector of observed data, and write $f(\mathbf{y} \mid \theta)$ to denote the sampling density of the data given the value of the parameter θ. The prior density for θ is denoted $g(\theta)$. From Bayes' theorem, it follows that the posterior density, denoted by $g(\theta \mid \mathbf{y})$, is given by

$$g(\theta|\mathbf{y}) = \frac{1}{c(\mathbf{y})} f(\mathbf{y}|\theta)g(\theta),$$

where $c(\mathbf{y})$ is a normalization constant which ensures that $g(\theta|\mathbf{y})$ is a probability density, i.e., $\int g(\theta|\mathbf{y})d\theta = 1$. In general, the value of $c(\mathbf{y})$ is not known and the mathematical form of $g(\theta|\mathbf{y})$ makes it difficult to summarize values of the marginal posterior distribution of the components of θ.

For example, suppose that we are interested in the marginal distribution of a single component of θ, say θ_i, obtained by integrating out all other components of θ from the joint posterior; that is, we might want to estimate

$$g(\theta_i|\mathbf{y}) = \int g(\theta|\mathbf{y})d\theta_1 \cdots d\theta_{i-1}d\theta_{i+1} \cdots d\theta_k.$$

In addition to obtaining the marginal density of θ_i, we might also wish to determine the posterior mean of θ_i, which we denote by μ_{θ_i}. To find this value, we must compute another integral,

$$\mu_{\theta_i} = \frac{1}{c(\mathbf{y})} \int \theta_i g(\theta \mid \mathbf{y})d\theta.$$

In fact, determining the joint posterior distribution itself usually requires the evaluation of a high-dimensional integral to simply obtain the normalizing constant, given by

$$c(\mathbf{y}) = \int f(\mathbf{y}|\theta)g(\theta)d\theta.$$

Likewise, the standard deviation of θ_i, defined as $\sigma_{\theta_i} = \sqrt{\mu_{\theta_i}^{(2)} - (\mu_{\theta_i})^2}$, requires the evaluation of another integral to determine the second moment of θ_i, $\mu_{\theta_i}^{(2)}$; that is,

$$\mu_{\theta_i}^{(2)} = \frac{1}{c(\mathbf{y})} \int \theta_i^2 f(\mathbf{y}|\theta)g(\theta)d\theta.$$

Besides posterior moments, other important quantities also require evaluating integrals. For example, the probability that θ_i falls in the set A is given by

$$Pr(\theta_i \in A) = \frac{1}{c(\mathbf{y})} \int I_A(\theta_i) f(\mathbf{y}|\theta) g(\theta) d\theta,$$

where $I_A(\theta_i)$ is the indicator function that is equal to 1 when θ_i is in the set A, and 0 otherwise. Also, the posterior predictive density, $f(y^*|y)$, of a future observation y^*, is obtained from the integral

$$f(y^*|\mathbf{y}) = \frac{1}{c(\mathbf{y})} \int f(y^*|\theta) f(\mathbf{y}|\theta) g(\theta) d\theta,$$

again requiring an ability to evaluate potentially high-dimensional integrals.

In general, we need to be able to compute integrals of the form

$$\frac{1}{c(\mathbf{y})} \int h(\theta) f(\mathbf{y}|\theta) g(\theta) d\theta \equiv \int h(\theta) g(\theta|\mathbf{y}) d\theta \qquad (2.1)$$

if we are to perform Bayesian inference.

2.2 An example

To illustrate different computational methods for summarizing posterior distributions, let us examine data collected in a study described in Dorn (1954) to assess the relationship between smoking and lung cancer. In this study, a sample of 86 lung-cancer patients and a sample of 86 controls were questioned about their smoking habits. The two groups were chosen to represent random samples from a subpopulation of lung-cancer patients and an otherwise similar population of cancer-free individuals. Table 2.1 provides the number of smokers and nonsmokers in each group.

The purpose of the study was to determine if there was a significant difference between the smoking habits in the two groups and to estimate the magnitude of this difference, should it exist.

Let p_L and p_C denote the population proportions of lung-cancer patients and controls who smoke. To measure the difference between these two proportions, we could simply examine the posterior distribution of $p_L - p_C$. However, the distribution of this quantity is likely to be highly skewed. This makes summarizing the distribution difficult, and it also makes asymptotic approximations to the sampling distribution of the MLE and the asymptotic normal approximation to the posterior distribution less precise. For that reason, it is common practice to transform the

	Cancer	Control
Smokers	83	72
Nonsmokers	3	14

TABLE 2.1. Data from a study to learn about the association of smoking and lung cancer.

parameters of interest and, instead, study the log odds-ratio α, defined as

$$\alpha = \log\left(\frac{p_L/(1-p_L)}{p_C/(1-p_C)}\right).$$

The log odds-ratio is 0 when the two proportions are the same and is symmetric about 0 in the sense that reversing the roles of p_L and p_C simply changes its sign. It also happens that the empirical estimate of the log odds-ratio based on the observed proportions in each population is approximately normally distributed in even moderately sized studies. For these reasons, the log odds-ratio provides a more convenient measure for the statistical analysis of the relative differences between two proportions.

Substantive questions that might be posed in terms of the true value of the log odds-ratio between p_L and p_C include the following:

- Are lung cancer patients more likely to smoke than controls? This hypothesis is equivalent to the statement that the log odds-ratio is positive; i.e., $\alpha > 0$. We assess the plausibility of this hypothesis by computing the posterior probability $\Pr(\alpha > 0|\text{data})$.
- What is the magnitude of the effect of smoking in contracting lung cancer? Assuming that lung-cancer patients are more likely to smoke than controls, we might then wish to estimate the magnitude of the effect of smoking on contracting lung cancer. This question might be addressed by evaluating the posterior distribution of the log odds-ratio α.

Both questions require the study of the posterior distribution of α. The data at hand are the observed proportions of lung cancer and cancer-free individuals who smoke. In the first group, 83 of 86 smoked, while in the second, 72 of 86 individuals smoked. From Chapter 1, if we assume that the observed numbers of smokers in each group are independent given the proportions p_L and p_C, then we know that the likelihood function for the proportions p_L and p_C of smokers is the product of binomial densities

$$L(p_L, p_C) = p_L^{83}(1-p_L)^3 p_C^{72}(1-p_C)^{14}, \qquad 0 < p_L, p_C < 1. \qquad (2.2)$$

To specify a prior for α, our parameter of interest, we might assume that the smoking probabilities p_L and p_C are independent a priori. In this case, we might then specify locally uniform priors on the components of α, $\log(p_L/(1-p_L))$ and $\log(p_C/(1-p_C))$. If we make this assumption, that the priors on $\log(p_L/(1-p_L))$ and $\log(p_C/(1-p_C))$ are uniform, then by transforming back to the probability scale, we find that the induced priors on p_L and p_C are proportional to $[p_L(1-p_L)]^{-1}$ and $[p_C(1-p_C)]^{-1}$. Using this prior for each proportion, the joint prior density for p_L and p_C is proportional to

$$g(p_L, p_C) = p_L^{-1}(1-p_L)^{-1} p_C^{-1}(1-p_C)^{-1}, \qquad 0 < p_L < 1, \qquad 0 < p_C < 1. \qquad (2.3)$$

Combining the likelihood (2.2) with the prior (2.3), the joint posterior density for the proportions is proportional to

$$g(p_L, p_C | \text{data}) \propto p_L^{83-1}(1-p_L)^{3-1} p_C^{72-1}(1-p_C)^{14-1} \qquad (2.4)$$

for $0 < p_L < 1$, and $0 < p_C < 1$. Although the marginal densities of p_L and p_C are recognizable from (2.4)—both are of the beta form—the posterior distribution of a function of these two proportions is not easily determined. Neither the sum nor the difference of two beta random variables has a standard form, nor does the log odds-ratio between two beta variables. Although analytic expressions for these functions of two proportions can be derived, the complicated form of the posterior density makes evaluation of posterior probabilities difficult.

To illustrate this difficulty, consider the distribution of the log odds-ratio, our parameter of interest. Defining the log odds-product as

$$\eta = \log\left(\frac{p_L}{1-p_L} \times \frac{p_C}{1-p_C}\right)$$

and applying a transformation of variables from (p_L, p_C) to (α, η), it follows that the joint posterior density of α and η is proportional to

$$g(\alpha, \eta | \text{data}) \propto \frac{\exp(83t_1)}{(1+\exp(t_1))^{86}} \frac{\exp(72t_2)}{(1+\exp(t_2))^{86}},$$

$$-\infty < \alpha < \infty, \qquad -\infty < \eta < \infty,$$

where $t_1 = (\alpha + \eta)/2$ and $t_2 = (\eta - \alpha)/2$. From this expression, we can, in theory, obtain the marginal posterior density of α by integrating out the "nuisance" parameter η. However, in practice, this integration must be performed numerically.

The joint posterior density of the log odds-ratio α and the log odds-product η is displayed in Figure 2.1. The small circle in the plot corresponds to the modal value, or the value of (α, η) for which the density achieves its maximum. The three contour lines in the graph correspond to locations where the density is equal to 10%, 1% and .1% of the maximum density value. Note that the joint density is skewed toward large values of both parameters.

2.3   Non-Simulation-Based Algorithms

2.3.1   The Multivariate normal approximation

In practice, the Bayesian version of the central limit theorem assures us that in regular models,[1] the posterior distribution of model parameters approaches a mul-

---

[1]By regular models, we essentially mean models in which the number of parameters do not increase with the number of observations, and for which an increasing amount of information is obtained for each parameter as the number of observations becomes large. Perhaps the most common example of "irregular" models are random effect models in which only a limited number of observations are available to estimate each random effect.

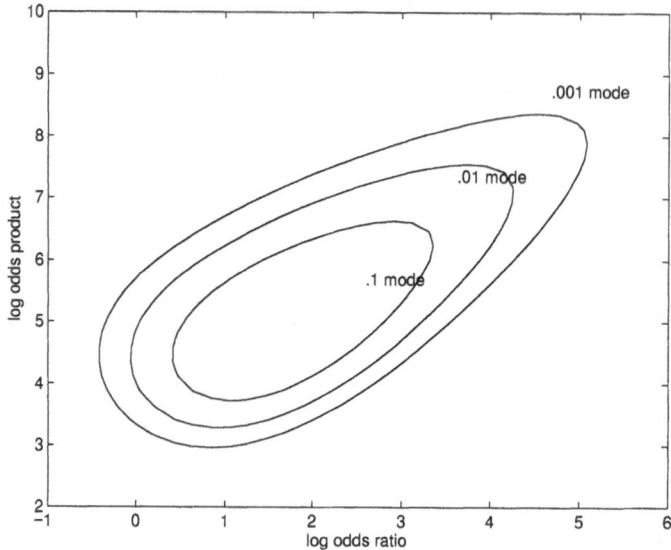

FIGURE 2.1. Contour plot of joint posterior density of log odds-ratio and log odds-product in smoking/cancer study.

tivariate normal distribution as the number of observations increases. It is therefore natural to approximate the posterior density in finite samples with a suitably centered and scaled normal density.

To find the appropriate mean and covariance matrix for the normal approximation to the posterior, we can expand the logarithm of the posterior density in a Taylor series expansion about its mode, $\tilde{\theta}$. Doing so yields

$$\log g(\theta|\text{data}) \approx \log g(\tilde{\theta}|\text{data}) + \frac{1}{2}(\theta - \tilde{\theta})' H(\theta - \hat{\theta}).$$

In this expansion, θ represents a column parameter vector, θ' is the transpose of θ, and H is the second derivative matrix of the logarithm of the posterior density evaluated at the mode. From this expression, the multivariate normal approximation to the posterior density follows from the fact that the logarithm of a multivariate normal density is exactly quadratic. The mean and covariance matrix of the corresponding multivariate normal density are $\tilde{\theta}$ and $(-H)^{-1}$, respectively.

This approximation is closely related to the large-sample approximation to the maximum likelihood estimator $\hat{\theta}$ discussed in Chapter 1. From a classical viewpoint, the sampling distribution of the MLE is approximately normal with mean θ and variance-covariance matrix $(-I)^{-1}$, where I is the second derivative matrix of the log-likelihood evaluated at the MLE. In the Bayesian setting, if a uniform prior $g(\theta)$ is used, then the posterior density and the likelihood are proportional, and $H = I$. Also, the posterior mode or MAP estimate is identical to the MLE. In this case, the Bayesian approximation is that the posterior density of θ is normal with mean $\hat{\theta}$ and variance-covariance matrix $(-I)^{-1}$. From either perspective, the distribution of $(\theta - \hat{\theta})$ is normal with mean 0 and variance-covariance matrix $(-I)^{-1}$. This equivalence means that the classical and Bayesian viewpoints will give the same approximate inferential procedure; that is, an approximate 95% classical confidence interval will be the same as the Bayesian 95% probability interval. Of course, the interpretation of the procedure (sampling based or probability based) differs between the two paradigms.

For the smoking/lung cancer study, $\theta = (\alpha, \eta)$ and the posterior mode is $\hat{\theta} = (1.68, 4.96)$. The approximate variance-covariance matrix is given by

$$(-H)^{-1} = \begin{pmatrix} .431 & .260 \\ .260 & .431 \end{pmatrix}.$$

Figure 2.2 illustrates the contour lines of the exact posterior density (solid line) and normal approximation to the posterior density (dotted line) for the smoking data. As expected, the normal approximation has an elliptical shape about the posterior mode. It resembles the exact posterior density in a region about the mode, but is less accurate in the extreme or tail portions of the distribution.

One important feature of the normal approximation is that it offers straightforward summaries of the posterior distribution. For example, the approximate marginal posterior density of the log odds-ratio α is normally distributed with mean $\mu = 1.68$ and standard deviation $\sigma = \sqrt{.431} = .66$. A 95% probability interval for α is also easy to obtain and is given by

$$(1.68 - z_{.025} \times .66, 1.68 + z_{.025} \times .66) = (.40, 2.97).$$

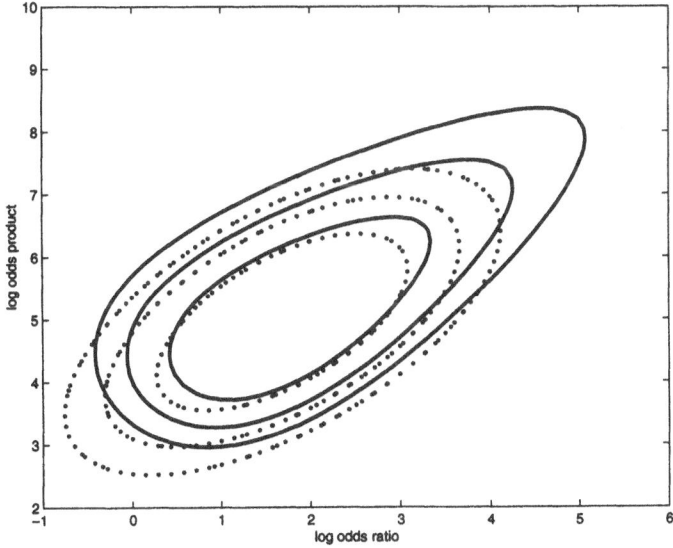

FIGURE 2.2. Contour plots of exact posterior density (solid line) and normal approximation (dotted line) for smoking/cancer study.

Similarly, the probability that the log odds-ratio is larger than 0 is approximated by the normal probability

$$\Pr(\alpha > 0) = \Pr\left(Z > \frac{0 - 1.68}{.66}\right) = .995.$$

Finally, the normal approximation to a posterior density also provides a useful mechanism for estimating the normalizing constant associated with the posterior. As will be seen in Chapter 3, the normalizing constant of the posterior distribution plays an important role in Bayesian model testing and is needed in the computation of Bayes factors. Letting $\phi(\theta; \mu, C)$ denote a multivariate normal density with mean μ and covariance matrix C, we can obtain an estimate of the normalizing constant $c(\mathbf{y})$ by equating the value of the posterior density at an arbitrary point θ to the approximating normal density according to

$$\phi(\theta; \mu, C) = \frac{1}{c(\mathbf{y})} f(\mathbf{y}|\theta)g(\theta).$$

A good choice of θ in this expression is usually provided by the estimated posterior mean $\hat{\theta}$. With this substitution, the normalizing constant is estimated by

$$c(\mathbf{y}) \approx \frac{f(\hat{\theta}|\mathbf{y})g(\hat{\theta})}{\phi(\hat{\theta}; \mu, \mathbf{C})},$$

where μ and C are estimated respectively by $\hat{\theta}$ and $(-H)^{-1}$.

Other methods for estimating a normalizing constant (or the marginal likelihood of the data) are described by Chib (1995), DiCiccio, Kass, Raftery and Wasserman (1997), Gelman and Meng (1998), Johnson (1998), and Meng and Wong (1996).

2.3.2   Grid integration

The normal approximation to the posterior density is based only on properties of first- and second-order derivatives of the posterior density evaluated at its mode. Clearly, more accurate computational algorithms can be developed by examining the shape of the posterior over a larger region. One such approach, adaptive quadrature, is based on approximating posterior integrals by sums of the form

$$\int f(\mathbf{y}|\theta)g(\theta)d\theta \approx \sum_{j=1}^{k} w^{(j)} f(\mathbf{y}|\theta^{(j)})g(\theta^{(j)}), \tag{2.5}$$

where $\theta^{(1)}, \ldots, \theta^{(k)}$ is a grid on which the posterior density is evaluated and $w^{(1)}, \ldots, w^{(k)}$ are weights assigned to each grid point. The location of the grid points and the assigned weights can be determined by Gauss-Hermite quadrature rules, in conjunction with estimates of the posterior mean and standard deviation. Gauss-Hermite quadrature rules are efficient when the posterior density is approximately proportional to the product of a normal density and a polynomial of low degree. Generally, if k grid points are used in Gauss-Hermite quadrature, the approximation (2.5) will be exact for integrals of products of polynomials of degree $2k - 1$ and a standard normal density.

To implement Gauss-Hermite quadrature for arbitrary posterior densities, it is necessary to center the quadrature points on the posterior mean and to scale the quadrature points by the posterior variance-covariance matrix. Quadrature points may be found in any number of books on numerical analysis, including *Handbook of Mathematical Functions*. Unfortunately, the parameterization used to represent quadrature formulas in most numerical analysis books is generally based on the error function rather than the standard normal distribution, and so it is necessary to convert from the "engineers" parameterization to the more appropriate statistical parameterization. Because Gauss-Hermite quadrature formulas are most often presented in the error function parameterization, we assume that parameterization of the quadrature points and weights in the discussion that follows. Specifically, we assume that the Gauss-Hermite quadrature formulas are presented for integrals of the form

$$\int_{-\infty}^{\infty} f(t)\,dt = \sum_{i=1}^{n} w_i \exp(x_i^2) f(x_i) \tag{2.6}$$

or, equivalently,

$$\int_{-\infty}^{\infty} \exp(-t^2) h(t)\,dt = \sum_{i=1}^{n} w_i h(x_i).$$

With this convention, we can approximate an integral of the form

$$\int_{-\infty}^{\infty} g(t)\,dt$$

by rescaling (2.6) to account for the nonzero mean and nonunit variance of the integrand to obtain

$$\sum_{i=1}^{n} \sqrt{2}\sigma w_i \exp(x_i^2) g(\sqrt{2}\sigma x_i + \mu). \tag{2.7}$$

In (2.7), $g(t)$ represents either a (possibly unnormalized) Gaussian density with mean μ and variance σ^2, or the product of a polynomial-like function and such a Gaussian density.

Although the mean and covariance matrix are usually not known a priori, the Gauss-Hermite quadrature scheme can be used in an iterative fashion to obtain both accurate estimates of these quantities and accurate estimates of the integrand. In other words, (2.7) can be applied iteratively to estimate μ and σ^2 along with the primary function of interest.

In higher dimensions, Gauss-Hermite quadrature formulas can be constructed by forming a grid of weights and quadrature points. For a d-dimensional parameter θ, the d-dimensional grid points may be written

$$(x_{i_1}, x_{i_2}, \ldots, x_{i_d}) \quad \text{for} \quad i_1, i_2, \ldots, i_d = 1, 2, \ldots, n.$$

The weights corresponding to each of these grid points are $w_{i_1}, w_{i_2}, \cdots, w_{i_d}$. Thus, a d-dimensional integral requires n^d function evaluations. If the mean and covariance matrix of the multivariate normal density appearing in the integrand are μ

and Σ, respectively, then the generalization of (2.7) to d dimensions is

$$\sum_{i_1=1}^{n}\sum_{i_2=1}^{n}\cdots\sum_{i_d=1}^{n}2^{d/2}|\Sigma|^{1/2}w_{i_1}w_{i_2}\cdots w_{i_d}\exp(\mathbf{x}_i'\mathbf{x}_i)g(\sqrt{2}\Sigma^{1/2}\mathbf{x}_i+\mu), \qquad (2.8)$$

where $\mathbf{x}_i' = (x_{i_1}, \ldots, x_{i_d})$. Here, $\Sigma^{1/2}$ denotes the symmetric square root of Σ. Technical details concerning the implementation of this algorithm can be found in the appendix located at the end of this chapter.

Figure 2.3 illustrates the application of this iterative scheme to the posterior distribution of (α, η) in the smoking example. Initial estimates of the means and standard deviations of α and η were -2 and .5, and 1 and .5, respectively. In addition, the two parameters were initially assumed to be uncorrelated. A 10×10 Gauss-Hermite grid of points was constructed from these estimates and is displayed in the top-left graph in Figure 2.3. We see from this plot that the initial estimates of the mean and standard deviations were poor, but the algorithm quickly begins adjusting in the next iteration. By the seventh iteration, the grid has stabilized near the true values of the posterior mean and covariance matrix. Note that the grid is skewed to account for the positive correlation between the two parameters. At the seventh iteration, the posterior mean and standard deviation of the log odds-ratio α based on this grid are estimated to be 1.82 and .70, respectively. Note that this posterior mean estimate is larger than the estimate based on the normal approximation. Much of the difference between the two estimates is due to the right skewness of the marginal posterior density of α.

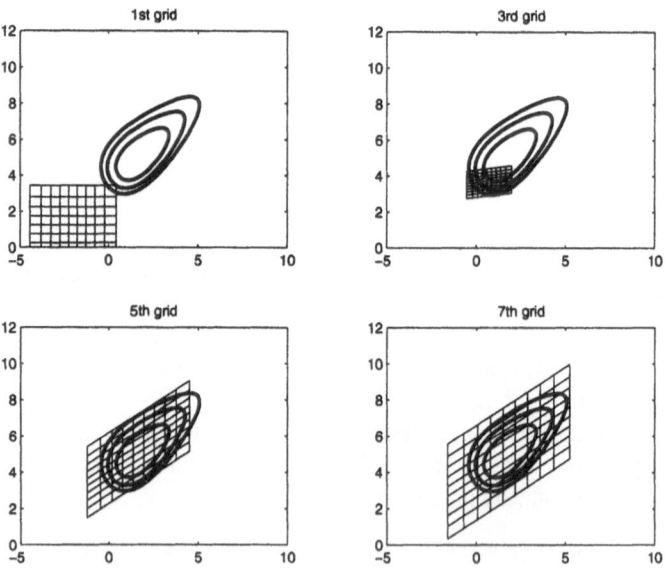

FIGURE 2.3. Illustration of a grid-based algorithm for summarizing a posterior density for smoking/cancer study. The exact posterior density and the grid of points is shown for four iterations of the algorithm.

2.3.3   Comments about the two computational methods

Both the normal approximation and Gauss-Hermite approximations are effective methods for integrating over a posterior density containing a small number of parameters. The normal approximation is based on the assumption that the posterior density is approximately quadratic on the logarithmic scale—or equivalently, that the posterior density is approximately Gaussian. If this assumption does not hold, either another approximation should be used or the parameter vector should be transformed to make the posterior more nearly normal. In part, that was why we chose to examine the log odds-ratio rather than the odds-ratio itself in the smoking example; the distribution of the log odds-ratio is approximately normal, while the distribution of the odds-ratio on the original scale is necessarily skewed, since it is truncated at 0. Generally, if the normal approximation is used, a d-dimensional parameter should be transformed so that it has support over d-dimensional Euclidean space.

Gauss-Hermite approximations also provide accurate summaries of many posterior integrals and are more accurate than the normal approximation since they are based on the shape of the posterior density at a comparatively large number of points. Gauss-Hermite quadrature rules are predicated on the assumption that the posterior density is approximately proportional to either a multivariate normal density or a multivariate normal density multiplied by a low-dimensional polynomial. However, the usefulness of Gauss-Hermite quadrature is limited to posterior densities defined over relatively low-dimensional parameter vectors. As mentioned above, this is so because these quadrature rules require evaluation of the posterior density on a grid containing, say, n values for each parameter component. When the parameter has d components, the total number of density evaluations is n^d, which grows exponentially with the dimension of the parameter vector. In many applications, this exponential growth of the grid makes Gauss-Hermite integration prohibitively expensive to implement.

2.4   Direct Simulation

Direct simulation methods for evaluating integrals of the form (2.1) are based on approximating (2.1) by the sum

$$\frac{1}{m} \sum_{j=1}^{m} h(\theta^{(j)}), \tag{2.9}$$

where $\theta^{(1)}, ..., \theta^{(m)}$ is an independent sequence of random variables drawn from the posterior distribution $g(\theta|y)$. In this section, we describe several methods for simulating the random variables $\theta^{(1)}, ..., \theta^{(m)}$ from the posterior distribution, and demonstrate the properties of (2.9). The method is illustrated using the data from the study of the relationship between smoking and lung cancer.

2.4.1   Simulating random variables

Fundamental to all of the simulation methods that we describe is the existence of a pseudo-random number generator that generates uniform random deviates from the interval $(0,1)$. Essentially all statistical software packages contain such a generator, but it is worthwhile to note that the variables generated are not truly random. However, the properties of these numbers are, for most purposes, indistinguishable from a independent sample of random values from the unit interval.

Inversion algorithm

Given a sequence of uniform random numbers, numerous algorithms have been developed for generating random variables from common univariate probability distributions. One generic method, called *inversion*, is based on inverting the *quantile* function (or inverse of the cumulative distribution function). Recall that for a continuous random variable θ with distribution G, the qth quantile θ_q is the number for which the probability that θ is less than or equal to θ_q is q:

$$G(\theta_q) = \Pr(\theta \le \theta_q) = \int_{-\infty}^{\theta_q} g(\theta)d\theta = q.$$

The quantile function $G^{-1}(q)$ is the inverse of the distribution function G — the function $G^{-1}(q)$ gives the quantile θ_q for an input probability of q. In implementing the inversion method, the quantile function $G^{-1}(q)$ is assumed to be known. To apply this method, we generate a sequence of uniform $(0,1)$ deviates $U^{(1)}, U^{(2)}, ..., U^{(m)}$ and take as our random sample $G^{-1}(U^{(1)}), ..., G^{-1}(U^{(m)})$. The values so obtained represent a random sample from the distribution function G.

To illustrate the inversion algorithm, suppose we want to generate a random sample from a beta density proportional to

$$g(p) = 5(1 - p)^4, \qquad 0 < p < 1.$$

The quantile function for this density is

$$G^{-1}(q) = 1 - (1 - q)^{1/5}, \qquad 0 < q < 1.$$

This quantile function can be used to find the median and quartiles of the beta density. The median is given by $G^{-1}(.5) = .129$, and the lower and upper quartiles are given by $G^{-1}(.25) = .056$ and $G^{-1}(.75) = .242$, respectively. Figure 2.4 (top) shows a histogram of a simulated sample $\{U^{(j)}\}$ of size 500 from the uniform $(0,1)$ distribution. Applying the quantile function to these uniform deviates results in the sampled beta density values depicted in the lower panel of Figure 2.4. For comparison, this plot also depicts the beta density $g(p)$; the shape of the histogram of simulated values appears to closely match the underlying probability density function.

Rejection algorithm

The inversion algorithm for simulating values from a probability distribution is useful in cases in which a simple expression exists for the quantile function. In

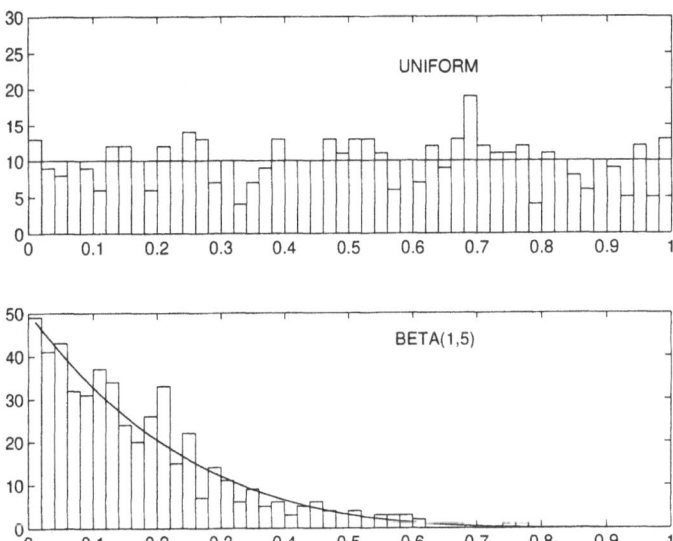

FIGURE 2.4. Demonstration of the inversion algorithm for simulating from a beta density. The top graph is a histogram of 1000 values from a uniform density and the bottom graph is a histogram of the beta simulated values.

cases in which the quantile function is not available, the rejection algorithm can sometimes be used.

Again, suppose that we are interested in obtaining a random sample from the probability density function $g(\theta)$. Suppose also that we know a second density function $f(\theta)$ that satisfies $g(\theta) \leq cf(\theta)$ for all θ and some positive constant c. Also, assume that generating random deviates with density $f(\theta)$ is easy.

Given the density f and constant c, random draws from g may be obtained by using the following rejection algorithm:

1. Simulate θ from $f(\theta)$, and U uniformly on $(0,1)$.
2. If $U < \frac{g(\theta)}{cf(\theta)}$, then accept θ as a draw from g. If not, reject θ and try again.

The algorithm is repeated until the desired sample size is obtained.

To illustrate the rejection algorithm, suppose we wish to sample from a beta density of the form

$$g(p) = 60p^3(1 - p)^2, \qquad 0 < p < 1.$$

Let $f(p) = 1$, $0 < p < 1$, denote a uniform density on the unit interval. The maximum value of $g(p)$ is 2.07, so this density is bounded above by $2.07f(p)$, as illustrated in the top panel of Figure 2.5. The rejection algorithm proceeds through the following steps:

1. Simulate p from f and U uniformly on $(0,1)$. (In this case, p and f have the same distribution.)
2. If $U < \frac{g(p)}{2.07f(p)}$, then accept p as a simulated value from g.

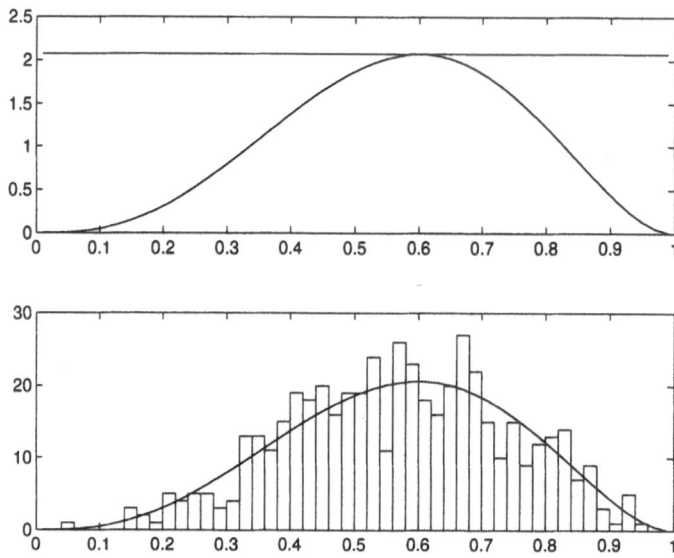

FIGURE 2.5. Demonstration of the rejection algorithm for simulating from a beta density. The top graph show the beta density and the bounding function and the bottom graph is a histogram of the beta simulated values.

To illustrate the algorithm, we repeated these steps 1000 times. Of the 1000 candidate values of p drawn, 497 were accepted, resulting in a random sample of that size. The bottom panel in Figure 2.5 depicts a histogram of this random sample. For comparison, the target density function g is also illustrated.

2.4.2   Inference based on simulated samples

Random samples from the posterior distribution can be used to explore most questions regarding the parameter of interest. Letting $\theta^{(1)}, \ldots, \theta^{(m)}$ denote such a sample, the shapes of the marginal densities for the components of θ can be explored by simple histogram or kernel density estimates of the simulated parameter values. The probability that θ or a function of θ belongs to a particular set can be approximated by the sample proportion of simulated values that belong to the same region. In addition, expectations taken with respect to the posterior distribution can be approximated using sample means of functions of the simulated values.

2.4.3   Inference for a binomial proportion

We now return to the student survey example of Chapter 1 and demonstrate the use of simulated samples for the purpose of performing inference for a binomial proportion. In that survey, we observed 5 successes (students who reported that expected grading policy had played no role in their decision to enroll in a course taken to satisfy a distributional requirement) out of 16 students surveyed. For simplicity, we assume a uniform prior for the binomial probability of success

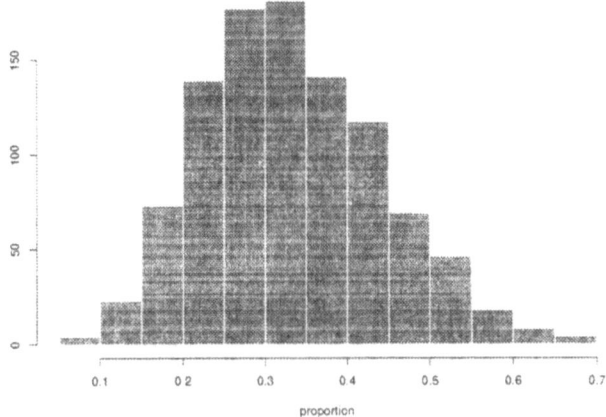

FIGURE 2.6. Histogram of simulated sample from a beta posterior density produced by direct simulation.

p. These data and prior lead to a beta posterior density (parameters 6 and 12) proportional to

$$g(p|\text{data}) \propto p^5(1-p)^{11}, \qquad 0 < p < 1.$$

Using the direct simulation techniques described in the last section, it is straightforward to obtain a random sample from this density; Figure 2.6 shows such a sample for 1000 simulated values from the posterior. Denote this sample by $p^{(1)}, \ldots, p^{(m)}$.

The different inferential questions discussed in Chapter 1 can all be addressed using the sampled values $p^{(1)}, \ldots, p^{(m)}$. For example, the mean of the posterior distribution, μ_p, can be estimated as the sample mean of the simulated proportion values:

$$\bar{p} = \frac{\sum_{j=1}^{m} p^{(j)}}{m}.$$

Likewise, the posterior standard deviation σ_p can be approximated by the sample standard deviation of the simulated values:

$$s_p = \sqrt{\frac{\sum_{j=1}^{m}(p^{(j)} - \bar{p})^2}{m-1}}.$$

The posterior median is estimated by the sample median of $p^{(1)}, \ldots, p^{(m)}$. The 5th, 25th, 75th, and 95th percentiles of the posterior distribution are approximated by the corresponding percentiles of the simulated values. The probability that the proportion p exceeds 1/2 is estimated by the proportion of simulated values that exceed .5.

	μ_p	σ_p	$p_{.05}$	$p_{.25}$	$p_{.50}$	$p_{.75}$	$p_{.95}$	$\Pr(p > .5)$
EXACT	.333	.108	.166	.255	.327	.405	.522	.072
SIMUL.	.332	.108	.170	.252	.322	.406	.527	.076
SE	.003							.008

TABLE 2.2. Exact and simulation-based posterior summaries of the density proportional to $p^5(1 - p)^{11}$.

2.4.4 Accuracy of posterior simulation computations

How accurate are the posterior summaries computed by simulation? Table 2.2 displays the simulation estimates of the posterior quantities discussed above and the corresponding estimates based on an essentially exact method for evaluating a beta distribution. Generally, we see close agreement between the two sets of values — the simulation estimates (using a sample size of 1000) appear accurate to the hundredths place.

Obviously, "exact" posterior summaries like those given in Table 2.2 will not usually be available, and so methods for assessing the accuracy of simulation output are needed. In general, we are interested in posterior expectations of the form $E(h(p|\text{data}))$, and our simulation estimate is $\bar{h} = \sum_{j=1}^{m} h(p^{(j)})/m$. From classical sampling theory, this estimate has associated standard error

$$se_{\bar{h}} = \sqrt{\frac{\sum_{j=1}^{m}(h(p^{(j)}) - \bar{h})^2}{(m-1)m}}.$$

For example, the estimate of the posterior mean of p is the sample mean $\bar{p}$. The estimated standard error of this estimate is

$$se_{\bar{p}} = \frac{s_p}{\sqrt{m}},$$

where s_p is the standard deviation of the simulated values $\{p^{(j)}\}$.

Similarly, an estimate of the posterior probability $q = \Pr(p \in A|\text{data})$ is given by the sample proportion of simulated values in A, $\bar{q} = \sum_{j=1}^{m} I(p^{(j)} \in A)/m$, where $I()$ is the indicator function. This proportion has associated standard error

$$se_{\bar{q}} = \sqrt{\frac{\bar{q}(1-\bar{q})}{m}}.$$

The standard errors of the posterior mean μ_p and posterior probability $P(p > .5)$ are shown in the "SE" row of Table 2.2. To illustrate the interpretation of these values, consider the computation of the posterior mean. The simulation estimate of the mean is .332, with an associated standard error of .003. Since the distribution of the simulation estimates is approximately normal, we believe that the exact value of the mean falls within two standard errors of the estimate; that is, we believe that μ_p lies in the interval $(.332 - 2(.003), .332 + 2(.003)) = (.326, .338)$ with 95% probability. Generally, the small size of the standard errors confirms the accuracy of the simulation estimates of the posterior mean and posterior probability.

2.4.5   Direct simulation for a multiparameter posterior: The composition method

In situations involving a multidimensional vector θ, it may be difficult to simulate directly from the posterior distribution on the entire parameter vector. However, in some cases, the simulation procedure can be simplified by dividing θ into two subcomponents θ_1 and θ_2, i.e., $\theta = (\theta_1, \theta_2)$. In such cases, the joint posterior density can be decomposed into the product

$$g(\theta) = g_1(\theta_1 | \text{data})g_2(\theta_2 | \theta_1, \text{data}).$$

If the component densities g_1 and g_2 have familiar functional forms and are easy to simulate, then the following method of composition may be used to simulate a value of θ:

- Draw θ_1 from the density g_1 and call the simulated value θ_1^*.
- Draw θ_2 from the density of g_2 conditional on $\theta_1 = \theta_1^*$. Let θ_2^* denote the simulated value.

The value $\theta^* = (\theta_1^*, \theta_2^*)$ then represents a draw from the joint distribution g.

2.4.6   Inference for a normal mean

The method of composition can be used to simulate from the joint posterior density of (μ, σ^2) in the normal error model of Chapter 1. Recall that the joint posterior density of the normal mean and variance was expressed as the product

$$g(\mu, \sigma^2 | \text{data}) = g(\sigma^2 | \text{data})g(\mu | \sigma^2, \text{data}),$$

where the marginal density of the variance, $g(\sigma^2 | x)$, is an inverse gamma density with parameters $(n - 1)/2$ and $S/2$, denoted by $\text{IG}((n - 1)/2, S/2)$, and the density of the mean, conditionally on the variance, $g(\mu | \sigma^2, x)$, was normally distributed with mean $\bar{x}$ and variance σ^2/n, respectively. To simulate a single value for (μ^*, σ^{2*}), the composition method may thus be implemented in the following way:

1. Generate σ^{2*} from an $\text{IG}((n - 1)/2, S/2)$ density.
2. Using the value σ^{2*} generated in Step 1, simulate μ^* from a $N(\bar{x}, \sigma^{2*}/n)$ density.

2.4.7   Direct simulation for a multiparameter posterior with independent components

Suppose that the parameter vector θ is subdivided into r components $\theta = (\theta_1, ..., \theta_r)$, and the components have independent posterior distributions. In this case, one can factor the joint density of θ as a product of the marginal densities of the components:

$$g(\theta) = \prod_{i=1}^{r} g(\theta_i).$$

Assume further that values can be simulated from each of the marginal densities $g(\theta_i), i = 1, ..., r$. Then if θ_i^* represents a simulated value from $g(\theta_i)$, $\theta^* = (\theta_1^*, ..., \theta_r^*)$ will be represent a random draw from the joint posterior distribution of θ.

2.4.8   Smoking example (continued)

Direct simulation can be used to effectively sample from the posterior distribution of the log odds-ratio in the smoking/lung cancer example. Rather than working with the posterior distribution of the log odds-ratio and the log odds-product, it is convenient to sample from the joint posterior distribution of the proportions of lung-cancer patients and control patients who smoke, p_L and p_C, respectively, and to then transform the sampled values to the (α, η) scale. The proportions have independent beta densities, with p_L distributed as a beta(83, 3) random variable and p_C distributed as a beta(72, 14) random variable. Thus, we can simulate from the posterior distribution of the log odds-ratio by sampling $p_L^{(1)}, ..., p_L^{(m)}$ from a beta(83, 3) density and sampling $p_C^{(1)}, ..., p_C^{(m)}$ from a beta(72, 14) density. Once the joint distribution of the two proportions has been sampled, it is straightforward to simulate the distribution of any function of the proportions. Specifically, a sample from the joint posterior distribution of the log odds-ratio, α, and the log odds-product, η, can be drawn by computing these functions for each pair of simulated values $\{(p_L^{(i)}, p_C^{(i)})\}$. A scatterplot of $m = 5000$ simulated values of the posterior of (α, η) computed using this direct simulation is displayed in Figure 2.7. A contour plot of the exact posterior density is also shown. Note that the shape of the scatterplot is similar to the shape of the contour plot, and most of the simulated values fall within the first contour (values of the density larger than 10% of the mode).

Once sample values of the the bivariate distribution of (α, η) have been simulated, inference about the marginal posterior density of the log odds-ratio can be performed by focusing on the values $\{\alpha^{(i)}\}$. Figure 2.8 displays a histogram of this simulated sample. The normal curve placed on top of the histogram is based on the normal approximation. It is clear from this graph that the "exact" posterior density constructed by simulation differs significantly from the one based on the normal approximation. The exact density is shifted to the right, so the difference between the two densities will be most noticeable in the computation of tail probabilities.

Table 2.3 provides summary statistics for the posterior distribution of α using the normal approximation, grid-based and direct simulation methods. Note that the grid-based and direct simulation methods yield posterior moment values that are identical to the second decimal. However, there are significant differences

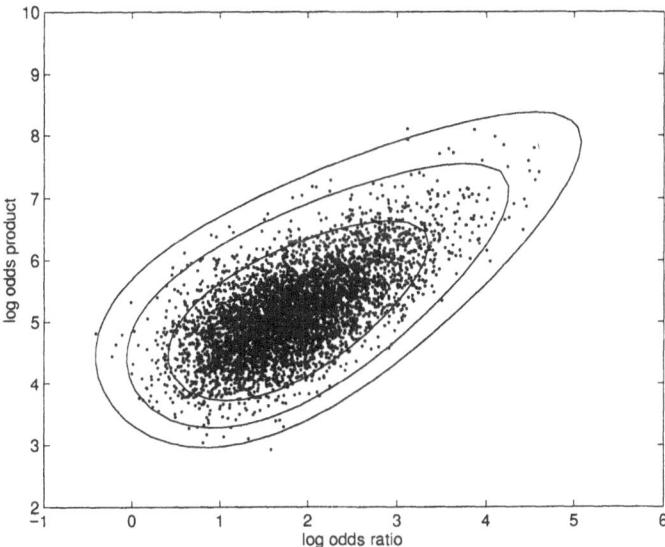

FIGURE 2.7. Scatterplot of simulated sample from the posterior density of the log odds-ratio and log odds-product produced by direct simulation. Contour graph of the exact posterior density is also shown.

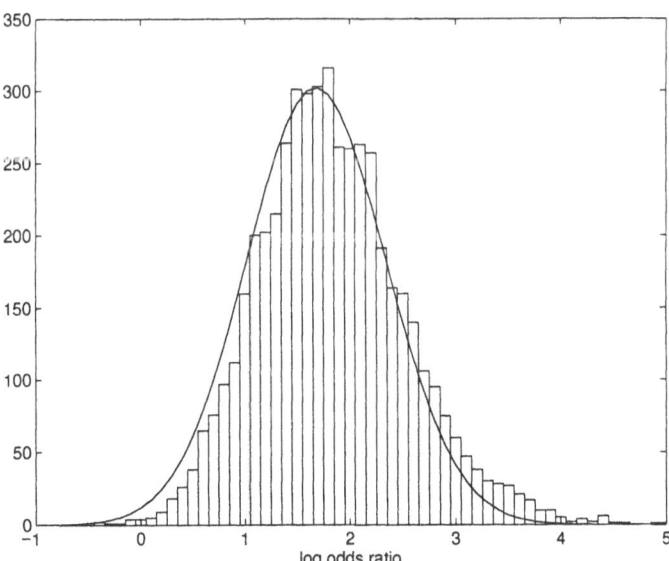

FIGURE 2.8. Histogram of simulated sample from the posterior density of the log odds-ratio produced by direct simulation. The approximate posterior density using the normal approximation is also shown.

Method	E(α)	SD(α)	95% interval	Pr($\alpha > 0$)
Normal approximation	1.68	.66	(.40, 2.97)	.995
Grid-based	1.82	.70		
Direct simulation	1.82	.70	(.60, 3.32)	.999
Metropolis	1.87	.68	(.74, 3.44)	.9996

TABLE 2.3. Posterior mean, posterior standard deviation, interval estimate, and tail probability of α computed using four computational methods.

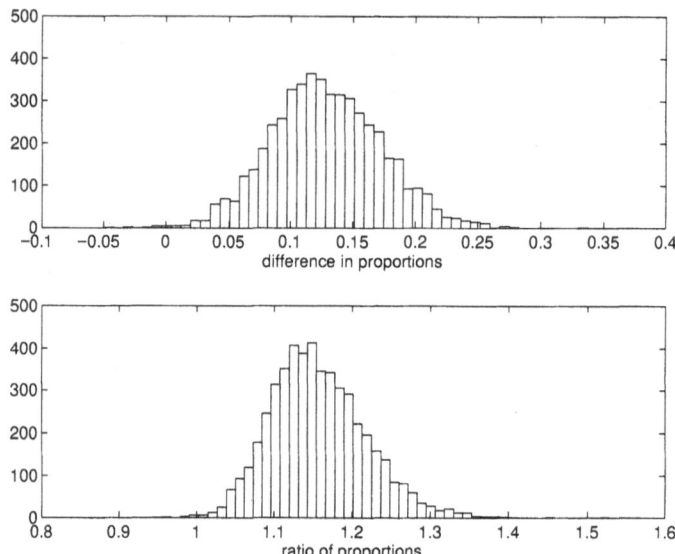

FIGURE 2.9. Histograms of simulated sample from the posterior density of the difference in proportions (top) and the ratio of proportions (bottom) using direct simulation.

between the normal approximation and simulation estimates, particularly for the 95% interval estimate for α and the probability that the log odds-ratio exceeds 0.

One attractive feature of direct simulation is that inference about any function of p_L and p_C can be performed directly using the set of sampled proportion values. For example, it is straightforward to obtain a histogram estimate of the difference in proportions $d = p_L - p_C$. Alternately, it may useful to compare the groups using the ratio of proportions $r = p_L/p_C$. Figure 2.9 shows histograms of the simulated values of the difference d (top plot) and the ratio r (bottom plot). As in the analyses above, we can summarize the marginal posterior samples of these measures of association by the computing sample means, standard deviations, and quantiles.

2.5 Markov Chain Monte Carlo

2.5.1 Introduction

Direct simulation can be used to summarize a one-parameter posterior distribution in the case where the density has a familiar functional form. For example, there are efficient routines for simulating normal, beta, gamma, and t random variables, and these routines can easily be used to summarize the posterior distribution of quantities derived from one of these standard families. In the multiparameter case, when the components of θ have independent posterior distributions, values for θ can be generated by independently drawing random variables for the component posterior distributions. Furthermore, the previous section illustrated the use of the composition method through which samples of random vectors could be generated by simulating, in turn, from the product of conditional densities.

Unfortunately, direct simulation applies only in models in which the posterior distribution has a familiar form. In many cases, the posterior distribution does not have such a form, and so alternative methods are required. In this section, we introduce a general class of methods for simulating from arbitrary, possibly high-dimensional posterior distributions. This class of algorithms, called collectively Markov chain Monte Carlo, or MCMC for short, has been successfully applied toward simulating from the posterior distribution in thousands of applications since their introduction to the statistical community in the early 1990s.

In the direct simulation methods described above, we were able to obtain independent draws from the posterior distribution. In MCMC algorithms, we, instead, obtain a correlated sequence of random variables in which the jth value in the sequence, say $\theta^{(j)}$, is sampled from a probability distribution dependent on the previous value, $\theta^{(j-1)}$. The exact distribution of $\{\theta^{(j)}\}$ is generally not known, although under conditions that are generally satisfied, the distribution of each iterate in the sequence of sampled values, $\theta^{(j)}$, does converge to the true posterior distribution as j becomes large. Thus, if a large number of sample updates are performed, the last group of sampled values in the sequence, say $\theta^{(K)}$, $\theta^{(K+1)}$, ..., $\theta^{(K+m)}$, will approximate a dependent sample from the posterior distribution of interest.

Essentially, MCMC algorithms produce random walks over a probability distribution. By taking a sufficient number of steps in this random walk, the simulation algorithm visits various regions of the state space in proportion to their posterior probabilities. For inferential purposes, the iterates obtained in these random walks can be summarized much like an independent sample from the posterior distribution.

MCMC algorithms are specified through a sequence of conditional probability densities $P(\theta^{(j)}|\theta^{(j-1)})$ and a method for moving from $\theta^{(j-1)}$ to the $\theta^{(j)}$, for $j = 2, \ldots$. One simple method for accomplishing this is the Metropolis-Hastings algorithm, which we now describe.

2.5.2 *Metropolis-Hastings sampling*

Metropolis-Hastings (MH) algorithms provide a simple, generic prescription for obtaining successive iterates in a MCMC run. The basic steps of a MH algorithm follow. For ease of description, we assume that θ is parameterized so that each of its components is real-valued and that the chain is currently in state $\theta^{(j-1)}$.

The first step in the algorithm is to generate a candidate point, denoted θ^c. Often, the candidate point differs from the current value of the parameter in only one or two components; for example, in the normal means problem, we may alternate between updating our value of μ and the value of σ. A common method for generating the candidate value θ^c is to add a mean zero normal deviate to a single component of $\theta^{(j-1)}$. For scalar parameters, this means that the candidate value may be expressed as

$$\theta^c = \theta^{(j-1)} + sZ, \tag{2.10}$$

where Z is a standard normal deviate and s is an arbitrary constant. For continuous-valued components of the parameter vector, let $\alpha(\theta^c|\theta^{(j-1)})$ denote the *proposal density* used to generate θ^c from $\theta^{(j-1)}$. In (2.10), α is a normal density with mean $\theta^{(j-1)}$ and standard deviation s. For discrete-valued components of the parameter vector, α represents the probability mass function used to generate candidate points. The probability of moving from the candidate point back to the original value is denoted, in a similar way, by $\alpha(\theta^{(j-1)}|\theta^c)$.

In choosing a rule to determine the proposal densities for each component of the chain, it is essential that the resulting algorithm be capable of moving from any point in the parameter space to any other point in a finite number of moves and that iterates in the chain be acyclic. We also require that the rule used to specify the proposal density satisfy

$$0 < \frac{\alpha(\theta^c|\theta^{(j-1)})}{\alpha(\theta^{(j-1)}|\theta^c)} < \infty$$

for all values $\theta^{(j-1)}$ and θ^c for which either the numerator or denominator are nonzero.

Having generated a candidate point, we next compute an *acceptance probability*, which is the probability that the candidate value will be accepted as the next simulated value in the sequence, $\theta^{(j)}$. This acceptance probability, denoted here by PROB, is calculated as

$$\text{PROB} = \min\left(1, \frac{g(\theta^c|\text{data})\,\alpha(\theta^{(j-1)}|\theta^c)}{g(\theta^{(j-1)}|\text{data})\alpha(\theta^c|\theta^{(j-1)})}\right).$$

In this formula, the acceptance probability represents the product of the ratio of the posterior probabilities of the candidate and current parameter values, $g(\theta^c|\text{data})/g(\theta^{(j-1)}|\text{data})$, and the ratio of the proposal densities of the current and candidate point, $\alpha(\theta^{(j-1)}|\theta^c)/\alpha(\theta^c|\theta^{(j-1)})$. The first ratio encourages the algorithm to move to parameter values having high posterior probability, and the second ratio accounts for the fact that the proposal density might favor some values of the parameter over others. Note that if the proposal density is symmetric—that is, if $\alpha(\theta^{(j-1)}|\theta^c) = \alpha(\theta^c|\theta^{(j-1)})$—this second ratio need not be computed.

The third step in a MH algorithm is to accept or reject the candidate point with probability equal to PROB. To do so, we draw a uniform (0,1) random variable, U, and compare U to PROB. If $U < PROB$, then we accept the candidate value and set $\theta^{(j)} = \theta^c$. On the other hand, if $U \geq PROB$, then we reject the candidate value and set $\theta^{(j)} = \theta^{(j-1)}$.

To illustrate the MH algorithm, let us first consider a simple one-parameter model for a proportion. In this stylized example, suppose that a company is considering production of an "improved" version of an old product and they wish to assess the market impact of the new version relative to the old. More specifically, suppose that the company wishes to determine the proportion p of potential customers who will prefer the new product to the old. Furthermore, due to the nature of the improvement, they are certain that the improved version will be at least as popular as the current product, implying that $p \geq .5$. To gain more precise knowledge about the population value of p, the company selects a random sample of 20 customers and asks each customer which version of the product they prefer. Of the 20 customers surveyed, 12 prefer the new version.

One way to model the prior beliefs held by the management of the company concerning the proportion p is to assume that the prior density for p is uniform on the interval (0.5, 1.0). If this prior is adopted, it follows from our discussion of binomial likelihoods in Chapter 1 that the posterior density must be proportional to

$$g(p|\text{data}) \propto p^{12}(1-p)^8, \qquad .5 < p < 1.$$

This posterior has the form of a beta density truncated to the interval (0.5, 1). Although this density can be summarized by numerically integrating the beta function, for purposes of illustration we will base our inference on simulated samples from the posterior density.

In order to use the MH algorithm to sample from this posterior, we first transform p to the logistic scale to obtain the real-valued parameter $\theta = \log[(p-.5)/(1-p)]$. Although such a transformation is not necessary, it simplifies the specification of the proposal density by eliminating difficulties near the boundary of interval (0.5, 1.0). The posterior density of the transformed parameter is proportional to

$$g(\theta|\text{data}) \propto \frac{(.5 + e^\theta)^{12} e^\theta}{(1 + e^\theta)^{22}}, \qquad -\infty < \theta < \infty.$$

To get an estimate of the location and spread of the posterior density of θ, we use the Taylor series expansion of the log posterior density around the mode. The mode and associated standard deviation are given by $-.89$ and $.63$, respectively. Of course, any reasonable estimates of these parameters could also be used to start the algorithm.

Next, we initialize the sequence by setting $\theta^{(0)} = -1$ (close to the posterior mode) and define the proposal density to be a normal distribution centered on the current iterate of the chain and having standard deviation $s = .6$ (roughly equal

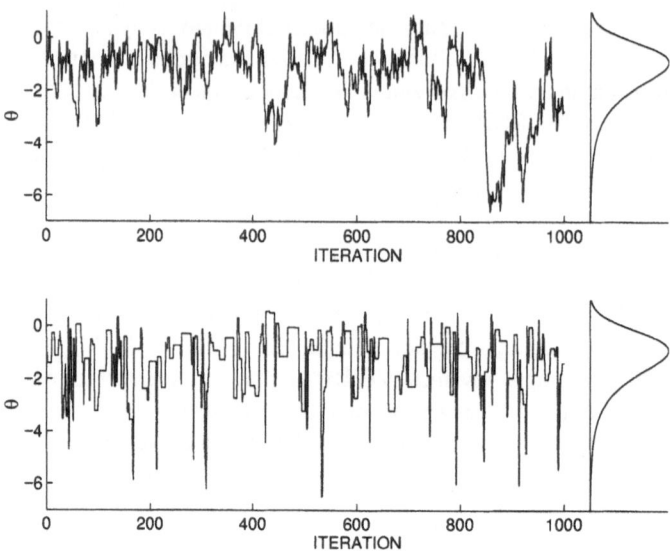

FIGURE 2.10. Sequence of simulated values using the Metropolis algorithm for the truncated beta example. The top graph uses a proposal density with $s = .6$ and the bottom graph uses a proposal density with $s = 6$. The exact posterior density of θ is graphed at the right-hand side of the figure.

to the posterior standard deviation of p). With these specifications, sampling from the posterior density using the MH algorithm proceeds using the steps outlined above.

One thousand simulated iterates from this chain are depicted in the top panel of Figure 2.10. For purposes of comparison, the true posterior density is plotted in the right-hand margin of the figure.

As Figure 2.10 illustrates, the MH algorithm appears to visit parameter values in proportion to their posterior probability. What is also clear from the figure is that consecutive values generated by the MH sampling scheme are correlated. This correlation between successive iterates in the chain is due, in part, to the fact that the standard deviation s used in the normal proposal density was relatively small compared to the range of plausible values in the posterior, and so most accepted updates in the sampler resulted in only small shifts from the previous update. Approximately 80% of candidate draws were accepted.

The acceptance rate of the algorithm can be lowered by increasing the width of the proposal density; that is, by increasing s. Hopefully, if s is not increased too much, the lower acceptance rate will be offset by larger jumps between accepted draws, and a net decrease in the correlation between iterates in the chain. To explore this possibility, the MH algorithm was rerun from the same starting value, but the proposal density was modified so that its standard deviation was 6. Results from this second run are displayed in the bottom plot in Figure 2.10. In this figure, there appears to be less dependence between consecutive values, although the algorithm appears more apt to "stick" or remain at the current sampled value. The acceptance rate for this version of the algorithm was only 20%. In practice, a common rule of thumb is to choose the proposal density so that the acceptance rate of the sampling scheme is between 25% and 40%. In this application, such a rate can be achieved by assigning s a value approximately equal to twice the estimated posterior standard deviation, or 1.2.

Of course, MH algorithms are not generally employed in such simple one parameter settings, so let us now examine a (slightly) more complicated two-parameter problem: simulation from the posterior density of the log odds-ratio and the log odds-product for the smoking/lung-cancer data. Recall that the normal approximation yielded estimates of the posterior mean equal to $\hat{\theta} = (1.68, 4.96)$, and posterior variance-covariance matrix

$$S = \begin{pmatrix} .431 & .260 \\ .260 & .431 \end{pmatrix}.$$

We use this approximate variance-covariance matrix in the construction of a bivariate MH sampler as follows:

1. Given the current value of θ in the simulation sequence, generate a candidate value according to the prescription

$$\theta^c = \theta + sZ,$$

where Z is multivariate normal with zero mean vector and variance-covariance matrix S.

2. Compute the acceptance probability

$$\text{PROB} = \min\left(1, \frac{g(\theta^c|\text{data})}{g(\theta|\text{data})}\right).$$

Note that the proposal density α does not enter into the computation of PROB, since it is symmetric in the sense that the probability of drawing point a when at point b is the same as drawing point b when starting from point a.

3. If a simulated random variable U is smaller than PROB, accept θ^c as the new simulated value, otherwise stay at the current value θ.

Figure 2.11 illustrates the first four iterations of the above MH algorithm for $s = 1$. A contour plot of the exact posterior density is depicted with dotted lines. The simulation algorithm was initialized at the point $\theta = (2, 4)$, which is relatively close to the posterior mode. The current point (at $(2, 4)$ in the top left plot) is indicated by an "x" in each subfigure. Candidate values (indicated by "o") are simulated from a normal density with covariance matrix S and centered on the current values (the solid ellipses indicate a contour line from this bivariate normal proposal density). If the candidate value has a higher posterior density value than the current point, then it is accepted with probability 1. Otherwise, it is accepted with probability equal to PROB. In iteration 3 (lower left panel), the candidate value drawn landed well away from the center of the posterior and thus had a high probability of being rejected.

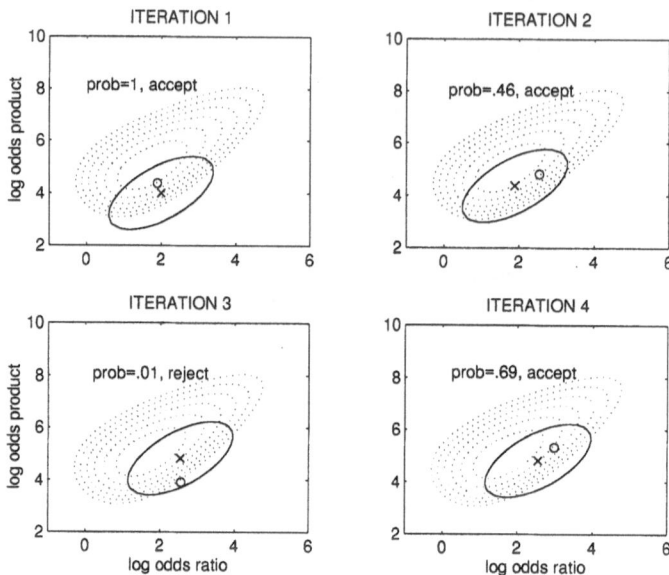

FIGURE 2.11. Demonstration of the Metropolis-Hastings algorithm for the posterior density of the log odds-ratio and the log odds-product. For each of four iterations, the graph shows the proposal and candidate values and the region where candidates can be generated. The exact posterior density is shown using a dotted line.

This MH algorithm was continued for 5000 iterations. A graph of the sequence of simulated values is shown in Figure 2.12. Note that there does not appear to be any trend in the sequence, which indicates that the collection of simulated values was approximately drawn from the posterior density of interest. We can summarize this sample of simulated values by means, standard deviations, and probabilities. In particular, the estimated mean, standard deviation, 95% probability interval, and tail probability were computed for the log odds-ratio α and are displayed in the last row of Table 2.3. Note that the values computed from the Metropolis-Hastings output agree quite well with the grid-based and direct simulation values obtained previously.

We leave it to the reader to compare this MH algorithm to a similar MH algorithm defined by updating the components of θ separately using normal proposal densities with standard deviations equal to $.66 = \sqrt{.431}$.

2.5.3   Gibbs sampling

The Metropolis-Hastings algorithm often provides an effective method for simulating from a posterior distribution of an unfamiliar form. However, the success of the method rests upon a reasonable choice of proposal density, which can be difficult in some cases. When inappropriate proposal densities are selected, one of two problems may arise. First, if the proposal density is too narrow, the MH algo-

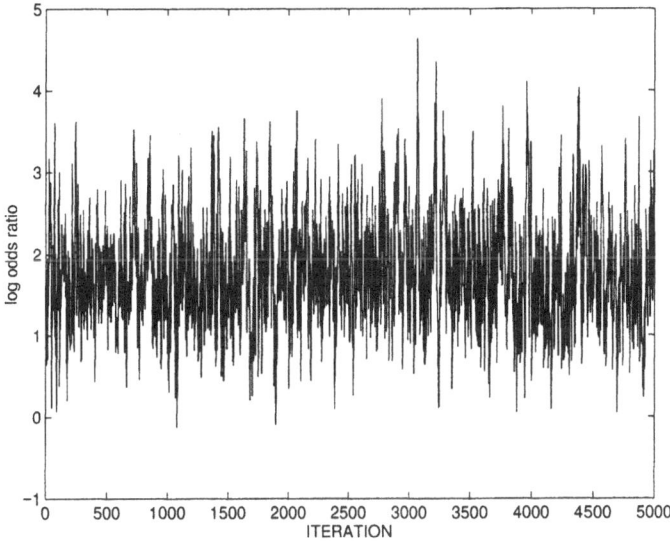

FIGURE 2.12. Graph of sequence of simulated values of the log odds-ratio using the Metropolis-Hastings algorithm.

rithm may spend all of its time in a limited region of the parameter space and may not be able to visit "distant" modes of the posterior in any reasonable number of updates (remember that MCMC algorithms provide samples from a posterior only in the limit—any run of finite length may not reach the equilibrium distribution!). In addition, overly narrow proposal densities result in high correlations between iterates of the chain, making the effective independent sample size quite small. On the other hand, if the proposal density is too broad, the chain may freeze in a single state for hundreds or even thousands of iterations, and little information can be gleaned from a run in which only 10 or 20 unique values are obtained.

Generally speaking, a better MCMC method for sampling from a posterior avoids the selection of a proposal density altogether by instead iteratively sampling from conditional posterior densities. MCMC algorithms that employ this strategy are often referred to as Gibbs samplers.

To describe a Gibbs sampler, suppose that the parameter vector θ can be partitioned into components $\theta = (\theta_1, \theta_2, \ldots, \theta_r)$, and denote the *full-conditional* posterior distributions by

$$p_1(\theta_1 | \theta_2, \ldots, \theta_r, \text{data})$$

$$p_2(\theta_2 | \theta_1, \theta_3, \ldots, \theta_r, \text{data})$$

$$\vdots$$

$$p_r(\theta_r | \theta_1, \ldots, \theta_{r-1}, \text{data}).$$

In the simplest case in which $r = 2$, the probability density p_1 represents the posterior density of component vector θ_1, conditional on the value of component θ_2. Likewise, p_2 is the conditional posterior density of θ_2 given θ_1. Although it may be difficult to simulate directly from the full parameter vector θ, it may be possible to generate simulated values from each of the full-conditional densities $p_1, p_2, \ldots, p_r$. This suggests the following MCMC simulation scheme. Let $\theta^{(j-1)} = (\theta_1^{(j-1)}, \theta_2^{(j-1)}, \ldots, \theta_r^{(j-1)})$ denote the simulated value at iteration $j - 1$. To get the next iterate $\theta^{(j)}$, we repeatedly subsample the components of $\theta^{(j-1)}$ according to the following steps:

1. Generate $\theta_1^{(j)}$ from $p_1(\theta_1 | \theta_2^{(j-1)}, \ldots, \theta_r^{(j-1)}, \text{data})$.
2. Generate $\theta_2^{(j)}$ from $p_2(\theta_2 | \theta_1^{(j)}, \theta_3^{(j-1)}, \ldots, \theta_r^{(j-1)}, \text{data})$. (Note that the value of θ_1 obtained in the previous step is used in this update.)

$$\vdots$$

r. Generate $\theta_r^{(j)}$ from $p_r(\theta_r | \theta_1^{(j)}, \ldots, \theta_{r-1}^{(j)}, \text{data})$.

At the completion of these steps, the vector $\theta^{(j)} = (\theta_1^{(j)}, \theta_2^{(j)}, \ldots, \theta_r^{(j)})$ provides the simulated value of θ at the jth iteration of sampling.

The r steps of this Gibbs sampling scheme constitute one iteration of the simulation method. Under suitable regularity conditions, the distribution of the random variate $\theta^{(j)}$ will converge to the posterior distribution of interest. In many applications, approximate convergence occurs after a small number of iterations, and the entire sequence $\{\theta^{(j)}\}$ may, for most purposes, be treated as a simulated sample from the posterior density $p(\theta | \text{data})$.

The Gibbs sampler is a particularly effective method for summarizing posterior distributions when there is missing or censored data. For example, consider the UV index data introduced in Chapter 1. In the original version of that example, we had 13 data points representing noontime UV measurements made on October Sundays in Los Angeles. However, suppose that after taking the UV measurements, it was later discovered that the measuring device had malfunctioned, recording all readings of 7 *or above* as 7. This means that the first observation of 7 might actually have represented a higher value, and we know only that it was at least 7. Data of this type are called censored data. Let us denote the observed values of the data by $y_1, \ldots, y_{13}$, which in this study were

$$7, 6, 5, 5, 3, 6, 5, 5, 3, 5, 5, 4, 4.$$

As in Chapter 1, we might still assume that the *actual*, uncensored UV measurements $x_1, \ldots, x_{13}$ are normally distributed with mean μ and standard deviation σ. Note that the actual and observed measurements are identical for the last 12 observations. However, due to censoring of the instrument, the "true" measurement for the first observation x_1 was not observed—we know only that it was at least 7. This example may be viewed as an inference problem with three unknowns: the two parameters of the normal distribution and the censored measurement. If the mean μ and the variance σ^2 are assigned the usual noninformative prior $g(\mu, \sigma^2) = 1/\sigma^2$, and x_1 is also assigned a noninformative prior, it follows that the joint posterior density of all unknowns has the form

$$g(\mu, \sigma^2, x_1 | \text{data}) = \frac{1}{\sigma^2} \prod_{i=2}^{12} \frac{1}{\sqrt{2\pi\sigma^2}} \exp\left\{-\frac{1}{2\sigma^2}(y_i - \mu)^2\right\},$$

$$\times \frac{1}{\sqrt{2\pi\sigma^2}} \exp\left\{-\frac{1}{2\sigma^2}(x_1 - \mu)^2\right\},$$

$$-\infty < \mu < \infty, \quad \sigma^2 > 0, \quad x_1 \geq 7.$$

To sample from the posterior distribution in this censored data context, we partition the vector of unknown quantities into $\theta = (\theta_1, \theta_2)$, where $\theta_1 = (\mu, \sigma^2)$ and $\theta_2 = x_1$. The full conditional distributions of the two component vectors have familiar forms and are easy to sample. The conditional posterior distribution of θ_1, given θ_2, simply reduces to sampling a normal mean and variance given a (uncensored) set of observations and can be accomplished using the normal/inverse-gamma distribution sampling scheme described earlier.

The distribution of θ_2 given θ_1 is only slightly more complex. If the parameters of the normal distribution are known, then the actual value of the censored observation is independently distributed according to a truncated normal distribution.

To specify the Gibbs sampling algorithm using these conditionals, we first initialize μ and σ^2, perhaps using estimates obtained by ignoring the censoring mechanism. One iteration of the Gibbs sampler is then comprised of the following steps:

1. Given current values of μ and σ^2, simulate a value of the censored observation x_1 from a normal $N(\mu, \sigma^2)$ distribution, truncated below by the censoring value 7.
2. Use the simulated values of the actual data generated in the above step to form a complete dataset $x_1, \ldots, x_{13}$. Simulate new values of the normal parameters μ and σ^2 from the normal/inverse gamma distribution.

This cycle is iterated a large number of times, storing the simulated values of the parameters at the end of each cycle. Graphical inspection of the sequence of the simulated values for μ and σ^2 indicates that the algorithm converges quickly, and so not too many iterations are required before the sampled values of μ and σ^2 might be regarded as samples from the posterior.

It is interesting to investigate the impact of censoring on inference about the normal mean. In particular, what would be the effect of the censoring if the investigator was unaware of the measuring defect and analyzed the data as if they were a random sample from a normal distribution? Figure 2.13 provides a partial answer to this question by comparing two posterior distributions of the mean μ. The plotted density is the "complete data" posterior density for μ obtained by ignoring the censoring mechanism. The histogram represents the exact posterior density which accounts for the censored observation. As we might expect, there are differences between the two distributions, although with only one censored observation, the differences are not extreme. The location of the true posterior density is larger than the location of the complete data posterior density—4.89 versus 4.84. In addition,

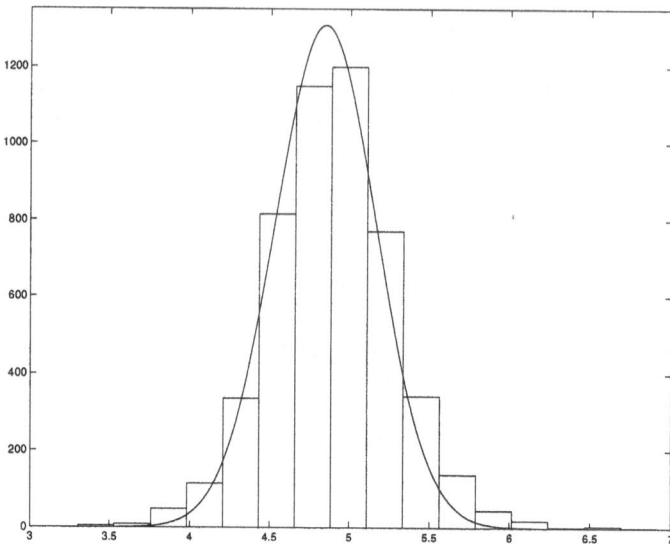

FIGURE 2.13. Histogram of simulated values of the normal mean for censored data example. The curve represents the posterior density of the mean that ignores the censoring mechanism.

the spread of the exact posterior is larger, reflecting the fact that the censored data contains less information than the complete data.

2.5.4   Output analysis

By running a MCMC algorithm, simulated values $\theta^{(1)}, ..., \theta^{(m)}$ that are each approximately distributed from the posterior distribution $g(\theta|\text{data})$ are obtained. For the MCMC algorithms that are used in this book, we know that the distribution of the simulated value at the jth iteration, $\theta^{(j)}$, will converge to a draw from the posterior distribution as j approaches infinity. Unfortunately, this theoretical result provides no practical guidance on how to decide if the simulated sample provides a reasonable approximation to the density $g(\theta|\text{data})$; that is, we do not know how long the chain must be run. Let us now briefly explore some of the important issues in interpreting MCMC output and describe graphical and numerical diagnostics for assessing convergence.

The first issue in understanding MCMC output is detecting the size of the *burn-in* period. Because of the fact that chains are initialized with values not actually drawn from the posterior distribution, the simulated values of θ obtained at the beginning of a MCMC run are not distributed from the posterior distribution. However, after some number of iterations have been performed (the burn-in period), the effect of the initial values wears off and the distribution of the new iterates approach the true posterior distribution. One way of estimating the length of this burn-in period is to examine *trace plots* of simulated values of a component (or some other function) of θ against the iteration number. (Examples of trace plots were shown in Figures 2.10 and 2.12.) Trace plots are particularly important when MCMC algorithms are initialized with parameter values that are far from the center of the posterior distribution. In such cases, the simulated values of θ in early iterations of the algorithm will drift toward the region of the parameter space where the posterior distribution is concentrated. An increasing or decreasing trend in the parameter values in the trace plot therefore indicates that the burn-in period is not over. Should such trends exist, it is important to discard the early portion of the chain, as the initial values obtained do not represent even approximate draws from the posterior.

A second concern in analyzing output from MCMC algorithms is the degree of *autocorrelation* in the sampled values. For both the MH and Gibbs sampling algorithms, the simulated value of θ at the $(j + 1)$st iteration is dependent on the simulated value at the jth iteration. If there is strong correlation between successive values in the chain, then two consecutive values provide only marginally more information about the posterior distribution than a single simulated value. A strong correlation between successive iterates indicates that the algorithm is "sticking" in a particular region of the parameter space and may take a long time to sample the entire distribution. In such cases, we say that the algorithm displays poor *mixing*.

A standard statistic for measuring the degree of dependence between successive draws in the chain is the autocorrelation. As its name suggests, the autocorrelation measures the correlation between sets of simulated values $\{\theta_i^{(j)}\}$ and $\{\theta_i^{(j+L)}\}$, where L is the lag or number of iterates separating the two sets of values. For a particular component or function of θ, one can compute the autocorrelation function as a function of differing values of the lag, L. For component i, the lag L autocorrelation may be estimated by

$$r_{iL} = \frac{m}{m - L} \frac{\sum_{i=1}^{m-L}(\theta_i - \bar{\theta})(\theta_{i+L} - \bar{\theta})}{\sum_{i=1}^{m}(\theta_i - \bar{\theta})^2},$$

where $\bar{\theta}$ is the mean of the simulated values. The value of the autocorrelation for lag 1 will almost always be positive for the MH and Gibbs sampling algorithms. However, if the chains are mixing adequately, the values of the autocorrelation will decrease to zero as the lag value is increased.

Another issue that arises in output analysis is the choice of the simulated sample size and the resulting accuracy of calculated posterior summaries. Because iterates in a MCMC algorithm are not independent, the expressions given in Section 2.4 cannot be used to compute standard errors of MCMC-based estimators. Instead, alternative estimates of simulation uncertainty have been developed. Perhaps the simplest of these is the method of *batch means*, which we now describe.

To illustrate the estimation of MCMC-sample uncertainty using batch means, suppose that we are interested in computing the posterior mean of a component of θ, say θ_i. An estimate of this posterior mean is given by the mean of the values in the simulated sample:

$$\bar{\theta}_i = \frac{\sum_{j=1}^{m} \theta_i^{(j)}}{m}.$$

To compute a standard error for this estimate, we subdivide the stream of simulated values $\theta_i^{(1)}, ..., \theta_i^{(m)}$ into b batches, each batch of size v, where $m = bv$. For each batch, we compute a sample mean; call the set of sample means $\bar{\theta}_i^1, ..., \bar{\theta}_i^b$. Suppose that the size of the batch v has been chosen large enough so that the autocorrelation (lag 1) in the sequence of batch means is small, say under .1. Then, the standard error of the estimate $\bar{\theta}_i$ can be approximated by the standard deviation of the batch means divided by the square root of the number of batches:

$$s_{\bar{\theta}_i}^B = \sqrt{\frac{\sum_{l=1}^b (\bar{\theta}_i^l - \bar{\theta}_i)^2}{(b-1)b}}.$$

This standard error is useful for determining the accuracy of posterior means that are computed in the simulation run. In the event that the standard error is too large, the MCMC algorithm should be rerun using a larger number of iterations.

Let us apply these output analysis techniques to the smoking/lung-cancer example. In Section 2.5.2, we illustrated the use of the MH algorithm to sample from the posterior distribution of the log odds-ratio and the log odds-product. We started the simulation algorithm at a point near the posterior mode, and the variance-covariance matrix of the multivariate normal proposal density was based on the Taylor series expansion of the log-posterior. Here, we use a MH algorithm using a poorly selected starting value $(-5, 10)$ which is far from the posterior mode. In addition, we use a normal proposal density with an identity variance-covariance matrix. This proposal density ignores the fact that the two parameters are positively correlated and the standard deviations of this density are larger than those based on the normal approximation.

The top panel of Figure 2.14 depicts a trace plot of the first 1000 simulated values of the log odds-ratio α using this MH algorithm. Although the algorithm was started some distance from the middle of the posterior, the simulated values quickly adjust and have moved to the center of the posterior distribution by iteration 20. So, even with a suboptimal sampler and poor starting values, the length of the burn-in for this low-dimensional problem is still quite short. Also, the simulated values do not appear to display any general increasing or decreasing pattern after iteration 20, indicating that the algorithm has approximately converged to the marginal posterior density of interest by this iteration.

The bottom panel of Figure 2.14 shows an autocorrelation plot of the 1000 simulated values of α. Although the lag 1 autocorrelation is approximately .9, the size of the autocorrelations decreases quickly as the lag increases to 20. This graph indicates that there is only modest autocorrelation in the MCMC run.

Next, suppose that we wish to bound the simulation-based uncertainty associated with the posterior mean of α based on this MCMC run. The sample mean of the 1000 simulated values was $\hat{\alpha} = 1.81$. We gauge the accuracy of this simulation-based computation by use of the batch means method. The calculations using this method are displayed in Table 2.4. The first row of the table presents the estimates using 1000 batches, each of size 1. The autocorrelation of the batch means in this case is .896, so in the second row of the table, we have increased the batch size by

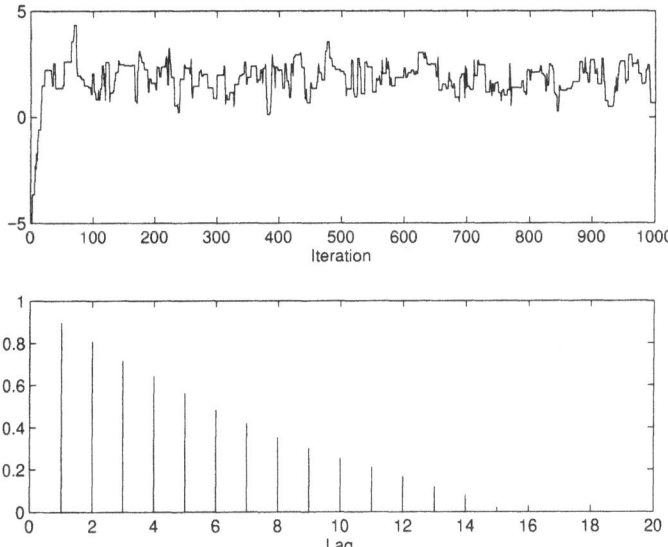

FIGURE 2.14. Trace plot (top) and autocorrelation plot (bottom) of a simulated stream of 1000 values of α using the MH algorithm for the smoking example.

Num. of batches	Batch size	Standard error	Lag 1 corr.
1000	1	.027	.896
500	2	.0378	.852
250	4	.0515	.742
125	8	.0692	.504
62	16	.0852	.077

TABLE 2.4. Illustration of the batch means method of assessing the standard error of the MCMC estimate of the posterior mean of α for the smoking example.

a factor of 2; there are then 500 batches of size 2. This process of increasing the batch size is continued until the lag 1 correlation of the batch means is under .1. From the table, we see that by dividing the run into 62 batches, each of size 16, the autocorrelation of the batch means is decreased to .077. These batch means are approximately independent, and the associated standard error of the estimate $\hat{\alpha}$ based on these batch means is .0852. Thus, we can be reasonably confident that the posterior mean is between $1.81 - 2 \times .09$ and $1.81 + 2 \times .09$.

2.6 A two-stage exchangeable model

In Section 1.3, a two-stage prior distribution was used to model the belief of exchangeability for a set of binomial probabilities. As we noted in that section, deriving analytic expressions for the marginal densities and posterior moments in that model was not possible. In this section, we illustrate simulation techniques for exploring the posterior distribution on that model's parameters using the composition method and the Metropolis-Hastings algorithm.

Recall that the model in question concerned the probabilities of death from stomach cancer for 20 cities in Iowa. The number of deaths in the ith city was denoted by y_i, for $i = 1, \ldots, 20$, and the total number of at risk individuals was denoted by n_i. We assumed that cancer deaths in each city were independently distributed according to a binomial distribution with probability p_i and denominator n_i. We also assumed a priori that our knowledge of the probabilities was exchangeable and reflected this prior knowledge by specifying that each probability was drawn from a common beta density. To summarize, we assumed the following:

- Given the values of p_i and n_i, y_i had independent binomial distributions with success probabilities p_i and n_i.
- Conditionally on the hyperparameters m and K, $p_1, \ldots, p_{20}$ were independently drawn from a common beta density with parameters Km and $K(1 - m)$.
- The hyperparameters m and K were assumed to be a priori independent, with m having a beta density with known parameters a_m and b_m, and K having a prior density proportional to $\frac{1}{(1+K)^2}$.

It follows that the joint prior density of the probabilities $\{p_i\}$ and unknown hyperparameters (m, K) is given by

$$g(\{p_i\}, m, K) \propto g(m, K) g(\{p_i\}|m, K)$$

$$= \frac{m^{a_m-1}(1 - m)^{b_m-1}}{(1 + K)^2} \prod_{i=1}^{20} \left(\frac{p_i^{Km-1}(1 - p_i)^{K(1-m)-1}}{B(Km, K(1 - m))} \right),$$

where $B()$ denotes the beta function defined as

$$B(a, b) = \frac{\Gamma(a)\Gamma(b)}{\Gamma(a + b)}.$$

The likelihood function is given by

$$L(\{p_i\}) = \prod_{i=1}^{20} p_i^{y_i}(1 - p_i)^{n_i-y_i}.$$

Combining the prior and likelihood function, the posterior density of the probabilities and hyperparameters can be expressed as

$$g(\{p_i\}, m, K|\text{data}) = g(m, K|\text{data}) g(\{p_i\}|m, K, \text{data}),$$

where the posterior density of the probabilities, conditional on the hyperparameters m and K, is given by

$$g(\{p_i\}|m, K, \text{data}) = \prod_{i=1}^{20} \frac{p_i^{Km+y_i-1}(1 - p_i)^{K(1-m)+n_i-y_i-1}}{B(Km + y_i, K(1 - m) + n_i - y_i)}. \qquad (2.11)$$

Because the conditional distributions of the p_i's have the form of beta densities, it is possible to analytically integrate these parameters out of the posterior to obtain the marginal posterior density of the hyperparameters,

$$g(m, K|\text{data}) \propto \frac{m^{a_m-1}(1-m)^{b_m-1}}{(1+K)^2} \prod_{i=1}^{20} \frac{B(Km+y_i, K(1-m)+n_i-y_i)}{B(Km, K(1-m))}.$$

(2.12)

The marginal posterior distribution of the hyperparameters m and K is intractable due to the beta functions appearing in the product.

Based on (2.11) and (2.12), we can design a MH scheme to sample from the joint posterior distribution over all quantities. The steps of such an algorithm may be summarized as follows:

1. Use the MH algorithm described in Section 2.5.2 to produce a sample from the joint posterior distribution of the hyperparameters m and K. As in previous examples, it is helpful to reparameterize so that the new model parameters are real-valued. To this end, take

$$\theta_1 = \log \frac{m}{1-m} \quad \text{and} \quad \theta_2 = \log K.$$

With this parameterization, given a current value of (θ_1, θ_2) in the chain, a candidate value (θ_1^c, θ_2^c) is generated as

$$\theta_1^c = \theta_1 + c_1 Z_1, \quad \theta_2^c = \theta_2 + c_2 Z_2,$$

where Z_1 and Z_2 are independent standard normal variates and c_1 and c_2 are known constants. The values of the constants are chosen by trial and error in pilot runs so that the acceptance rate of the algorithm falls in the interval (0.3, 0.5). Using this algorithm, we generate a large number, say 5000, of sampled values, denoted $(m^{(1)}, K^{(1)}), \ldots, (m^{(5,000)}, K^{(5,000)})$.

2. We use the composition method to simulate values of the cancer probabilities. For each simulated set of values of the hyperparameters, $(m^{(j)}, K^{(j)})$, we simulate the probabilities $p_1, \ldots, p_{20}$ independently from beta densities, where p_i is beta with parameters $y_i + K^{(j)}m^{(j)}$ and $n_i - y_i + K^{(j)}(1-m^{(j)})$. For $j = 1, \ldots, 5000$, denote the resulting sample values by $(m^{(j)}, K^{(j)}, p_1^{(j)}, \ldots, p_{20}^{(j)})$; these values represent a dependent sample from the joint posterior density of $(m, K, \{p_i\})$.

The simulated values $\{p_i^{(j)}, j = 1, \ldots, 5000\}$ are a dependent sample from the marginal posterior density of p_i and can be used for a variety of inferential purposes. For example, the posterior mean of p_i can be estimated from the simulated values according to

$$E(p_i|\text{data}) \approx \frac{\sum_{j=1}^{5000} p_i^{(j)}}{5000}.$$

This equation was used to generate the estimated posterior means displayed in Table 1.3. Alternatively, the simulated values can be used to estimate 90% probability intervals for the probabilities p_i by constructing the intervals of the form

$(p_i^{.05}, p_i^{.95})$, where p_i^q is the qth quantile of the set of simulated values from the marginal posterior density.

2.7 Further reading

The reader interested in additional details concerning MCMC algorithms can consult the recent text *Markov Chain Monte Carlo in Practice* edited by Gilks, Richardson, and Spiegelhalter and *Bayesian Computation Using Minitab* by Albert. Alternatively, we recommend several of the original papers on this topic, including Hastings (1970), Tanner and Wong (1987), Gelfand and Smith (1990), Gelfand, Hills, Racine-Poon, and Smith (1990), and Tierney (1994). Elementary discussions of Metropolis and Gibbs sampling are found in Chib and Greenberg (1995) and Casella and George (1992).

2.8 Appendix: Iterative implementation of Gauss-Hermite quadrature

In d dimensions, the iterative process for applying Gauss-Hermite quadrature to evaluate an integral of the form

$$\int_{-\infty}^{\infty} \cdots \int_{-\infty}^{\infty} h(\theta)g(\theta \mid y)d\theta_1 \cdots d\theta_d, \qquad (2.13)$$

where $g(\theta \mid y)$ denotes the (possibly unnormalized) posterior density, may be summarized in the following steps:

1. Choose an initial estimate of the posterior mean and variance, μ_0 and Σ_0, respectively.
2. Apply Gauss-Hermite quadrature rules to obtain the following values:

   a.

$$a = \sum_{i_1=1}^{n}\sum_{i_2=1}^{n}\cdots\sum_{i_d=1}^{n} 2^{d/2}\{|\Sigma_0|^{1/2}w_{i_1} w_{i_2} \cdots w_{i_d}$$
$$\times \exp(\mathbf{x}_i'\mathbf{x}_i)\, g(\sqrt{2}\Sigma_0^{1/2}\mathbf{x}_i + \mu_0)\},$$

   b. For $j = 1, \ldots, d$.

$$b_j = \sum_{i_1=1}^{n}\sum_{i_2=1}^{n}\cdots\sum_{i_d=1}^{n} 2^{d/2}\{|\Sigma_0|^{1/2}w_{i_1} w_{i_2} \cdots w_{i_d} \exp(\mathbf{x}_i'\mathbf{x}_i)$$
$$\times z_j\, g(\sqrt{2}\Sigma_0^{1/2}\mathbf{x}_i + \mu_0)\},$$

   where z_j denotes the jth component of the vector

$$\sqrt{2}\Sigma_0^{1/2}\mathbf{x}_i + \mu_0.$$

c. For $j = 1, \ldots, d$ and $k = j, \ldots, d$,

$$c_{jk} = \sum_{i_1=1}^{n} \sum_{i_2=1}^{n} \cdots \sum_{i_d=1}^{n} 2^{d/2} \{|\Sigma_0|^{1/2} w_{i_1} w_{i_2} \cdots w_{i_d} \exp(x_i' x_i)$$
$$\times z_j z_k g(\sqrt{2} \Sigma_0^{1/2} x_i + \mu_0)\},$$

3. Update the initial estimate of the mean and covariance matrix. Set the jth component of μ_0 to be

$$\mu_j = b_j / a$$

and the (j, k) element of Σ_0 to be

$$\sigma_{jk} = \sigma_{kj} = c_{jk}/a - \mu_j \mu_k, \qquad j \leq k.$$

4. Repeat Steps 2 and 3 until the estimates of μ_0 and Σ_0 stabilize.
5. Estimate the value of (2.13) by

$$\sum_{i_1=1}^{n} \sum_{i_2=1}^{n} \cdots \sum_{i_d=1}^{n} \{2^{d/2} |\Sigma_0|^{1/2} w_{i_1} w_{i_2} \cdots w_{i_d} \exp(x_i' x_i)$$
$$\times h(\sqrt{2} \Sigma_0^{1/2} x_i + \mu_0) g(\sqrt{2} \Sigma_0^{1/2} x_i + \mu_0)\}. \tag{2.14}$$

If $g(\cdot)$ represents the unnormalized posterior density and the integral corresponding to the normalized posterior is desired, divide (2.14) by the last value of a obtained in Step 2.

2.9 Exercises

1. (Estimating a log-odds with a normal prior.) Suppose y has a binomial distribution with parameters n and p, and we are interested in the log-odds value $\theta = \log\left(\frac{p}{1-p}\right)$. Our prior for θ is that $\theta \sim N(\mu, \sigma^2)$. It follows that the posterior density of θ is given, up to a proportionality constant, by

$$g(\theta|y) \propto \frac{\exp(y\theta)}{(1 + \exp(\theta))^n} \exp\left[\frac{-(\theta - \mu)^2}{2\sigma^2}\right].$$

More concretely, suppose we are interested in learning about the probability that a special coin lands heads when tossed. A priori, we believe that the coin is fair, so we assign θ a $N(0, .5)$ prior. We toss the coin $n = 5$ times and obtain $y = 5$ heads.

a. Using a normal approximation to the posterior density, compute the probability that the coin is biased toward heads (i.e., that θ is positive).
b. Use Gauss-Hermite quadrature to approximate the posterior mean and standard deviation. Compare these results to those obtained in (a) using the normal approximation.

c. Use a Metropolis-Hastings random-walk algorithm to simulate from the posterior density. In the algorithm, let s be equal to twice the approximate posterior standard deviation found in (a). Use the simulation output to approximate the posterior mean and standard deviation of θ, and the posterior probability that θ is positive. Compare your answers with those obtained in (a) and (b).

2. (Genetic Linkage Model from Rao (1973)) Suppose 197 animals are distributed into four categories with the following frequencies:

Category	1	2	3	4
Frequency	125	18	20	34

Assume that the probabilities of the four categories are given by the vector

$$\left(\frac{1}{2}+\frac{\theta}{4}, \frac{1}{4}(1-\theta), \frac{1}{4}(1-\theta), \frac{\theta}{4}\right),$$

where θ is an unknown parameter between 0 and 1. If θ is assigned a uniform prior, then the posterior density of θ is given by

$$g(\theta|data) \propto \left(\frac{1}{2}+\frac{\theta}{4}\right)^{125}\left(\frac{1}{4}(1-\theta)\right)^{18}\left(\frac{1}{4}(1-\theta)\right)^{20}\left(\frac{\theta}{4}\right)^{34},$$

where $0 < \theta < 1$. If θ is transformed to the real-valued logit $\eta = \log\frac{\theta}{1-\theta}$, then the posterior density of η can be written as

$$f(\eta|data) \propto \left(2+\frac{e^\eta}{1+e^\eta}\right)^{125}\frac{1}{(1+e^\eta)^{39}}\left(\frac{e^\eta}{1+e^\eta}\right)^{35}, \quad -\infty < \eta < \infty.$$

a. Using the normal approximation, find a 95% probability interval for η. Transform this interval to obtain a 95% probability interval for the original parameter of interest, θ.
b. Use a Metropolis-Hastings random-walk algorithm to simulate from the posterior density of η. (Choose the scale parameter s to be twice the approximate posterior standard deviation of η found in (a).) Compare the histogram of the simulated output of η with the normal approximation found in part (a). From the simulation output, find a 95% probability interval for the parameter of interest θ. Compare your interval with the probability interval computed in (a).

3. (Estimating the location and scale of a t population density.) Suppose that a random sample $x_1, \ldots, x_n$ is taken from a t density with 4 degrees of freedom, location parameter θ, and scale parameter σ:

$$f(x|\mu, \sigma) \propto \frac{1}{\sigma}\left(1+\frac{(x-\mu)^2}{4\sigma^2}\right)^{-5/2}.$$

Furthermore, assume that the vector of unknown parameters (μ, σ) is assigned the noninformative prior

$$g(\mu, \sigma) = \frac{1}{\sigma}, \qquad -\infty < \mu < \infty, \qquad \sigma > 0.$$

Suppose that we consider the joint posterior density of the location μ and the log scale $\eta = \log \sigma$. The joint density of these parameters is proportional to

$$g(\mu, \eta | \text{data}) \propto \prod_{i=1}^{n} \frac{1}{\exp(\eta)} \left(1 + \frac{(x_i - \mu)^2}{4 \exp(2\eta)} \right)^{-5/2},$$

where both μ and η are real-valued.

As an example of such data, consider an experiment by Charles Darwin (1876) in which pairs of seedlings of the same age were grown under similar conditions, except that one seedling in each pair was produced from cross-fertilization and one by self-fertilization. The difference in height (cross − self) was computed for each pair of seedlings after a fixed period of time. The differences in heights for the 15 pairs are given below:

6.1	−8.4	1.0	2.0	0.7	2.9	3.5	5.1
1.8	3.6	7.0	3.0	9.3	7.5	−6.0	

Find the posterior mean and standard deviation of μ using at least two different computational methods.

4. (Estimating the parameters of a Poisson/gamma density.) Suppose that $y_1, \ldots, y_n$ are a random sample from the Poisson/gamma density

$$f(y|a, b) = \frac{\Gamma(y + a)}{\Gamma(a) y!} \frac{b^a}{(b+1)^{y+a}},$$

where $a > 0$ and $b > 0$. This density is an appropriate model for observed counts which show more dispersion than predicted under a Poisson model. Suppose that (a, b) are assigned the noninformative prior proportional to $1/(ab)$. If we transform to the real-valued parameters $\theta_1 = \log a$ and $\theta_2 = \log b$, the posterior density is proportional to

$$g(\theta_1, \theta_2 | \text{data}) \propto \prod_{i=1}^{n} \frac{\Gamma(y_i + a)}{\Gamma(a) y_i!} \frac{b^a}{(b+1)^{y_i+a}},$$

where $a = \exp\{\theta_1\}$ and $b = \exp\{\theta_2\}$. Use this framework to model data collected by Gilchrist (1984), in which a series of 33 insect traps were set across sand dunes and the numbers of different insects caught over a fixed time were recorded. The number of insects of the taxa Staphylinoidea caught in the traps are shown below.

2	5	0	2	3	1	3	4	3	0	3
2	1	1	0	6	0	0	3	0	1	1
5	0	1	2	0	0	2	1	1	1	0

Using the normal approximation and Metropolis-Hastings algorithms, find 90% probability intervals for a and b. (In the use of the Metropolis-Hastings sampler, generate candidate values using the rule $\theta^c = \theta + sZ$, where $s = 2$, Z is $N(0, S)$, and S is the approximate variance-covariance matrix used in the normal approximation.)

5. (Comparing two Poisson rates.) Suppose that the number of "first" births to women during a month at a particular hospital has a Poisson distribution with parameter R. During a given year at a Swiss hospital (Walser, 1969), 66 births were recorded in January and 48 births were recorded in April. Is there strong evidence that the birth rate R_J during January exceeds the birth rate R_A of April?

    a. Assuming that the vector (R_J, R_A) has a noninformative prior proportional to $1/(R_J R_A)$, find the posterior density for R_J and R_A. Show that R_J and R_A have independent gamma distributions.

    b. Using direct simulation, simulate from the marginal posterior density of the ratio of rates R_J/R_A.

    c. Approximate the posterior probability that $R_J > R_A$ from the simulation output.

6. (The Behrens-Fisher problem.) Suppose that we observe two independent normal samples, the first distributed according to a $N(\mu_1, \sigma_1^2)$ distribution, and the second according to a $N(\mu_2, \sigma_2^2)$ distribution. Denote the first sample by $x_1, ..., x_m$ and the second sample by $y_1, ..., y_n$. Suppose further that the parameters $(\mu_1, \sigma_1^2, \mu_2, \sigma_2^2)$ are assigned the vague prior

$$g(\mu_1, \sigma_1^2, \mu_2, \sigma_2^2) = \frac{1}{\sigma_1^2 \sigma_2^2}.$$

    a. Find the posterior density. Show that the vectors (μ_1, σ_1^2) and (μ_2, σ_2^2) have independent posterior distributions.

    b. Using the method of composition, describe how to simulate from the joint posterior density of $(\mu_1, \sigma_1^2, \mu_2, \sigma_2^2)$.

    c. The following data from Manly (1991) give the mandible lengths in millimeters for 10 male and 10 female golden jackals in the collection of the British Museum. Using simulation, find the posterior density of the difference in mean mandible length between the sexes. Is there sufficient evidence to conclude that the males have a larger average?

Males									
120	107	110	116	114	111	113	117	114	112
Females									
110	111	107	108	110	105	107	106	111	111

7. (Stomach-cancer data revisited.) As an alternative model for the exchangeability of the stomach-cancer probabilities, we might assume that the parameters a and b were drawn from independent gamma distributions with parameters (c, λ) and (d, λ), respectively; that is, we might assume that

$$a \sim \Gamma(c, \lambda) \quad \text{and} \quad b \sim \Gamma(d, \lambda).$$

Under such an assumption, the variable $a/(a + b)$ has a beta density with parameters $c/(c + d)$.

a. Specify values for c, d, and λ and explain why those values might provide a reasonable model for this data.
b. Repeat the analysis of Section 2.6 for this prior model. In particular, design a MCMC algorithm to sample from the posterior distributions of the parameters p_i, a, and b.
c. Plot the predictive density for the probability of death from stomach cancer for a Iowa city not included in the study.

8. (Illustrating different computational methods for estimating a Poisson mean.) Exercise 1.5 examined the problem of estimating the breakdown rate R of a fleet of trucks. In that problem, the observed numbers of breakdowns of trucks on n consecutive days, $y_1, ..., y_n$, were assumed to be independently distributed from a Poisson distributions with rate parameter R. If R is assigned the noninformative prior $g(R) = 1/R$, then the posterior density on R is given by

$$g(R|\text{data}) \propto \exp\{-nR\}R^{s-1}, \quad R > 0,$$

where $s = \sum_{i=1}^{n} y_i$. In this case, $n = 5$ and $s = 11$. Now suppose that we are interested in two additional quantities: (1) a 90% probability interval for R and (2) the posterior probability that R exceeds 4.

a. Find "exact" values for the probability interval and $\Pr(R > 4)$.
b. To illustrate the use of the normal approximation described in Section 2.3, suppose that the positive parameter R is transformed to the real-valued parameter $\theta = \log R$. The posterior density of θ is given by

$$g(\theta|\text{data}) \propto \exp\{-ne^{\theta} + s\theta\}, \quad \infty < \theta < \infty.$$

Find a normal approximation to this posterior density. Use this approximation to compute the two quantities of interest above. Compare your answers with the exact answers computed in part (a).
c. Using the Metropolis-Hastings algorithm described in Section 2.5, simulate a sample of size 1000 from the posterior density of the transformed parameter θ. Start the algorithm at the posterior mode and generate candidate values $\theta^c = \theta + sZ$, where $s = 2$ and Z is normal with mean 0 and variance S, where S is estimated from the normal approximation in part (b). Again, use this simulated sample to compute the two quantities of interest. Compare your answers with the exact answers computed in part (a).

9. (Estimating normal parameters from rounded data.) Reconsider the UV index data introduced in Chapter 1. Suppose that the observed noontime UV measurements

$$7, 6, 5, 5, 3, 6, 5, 5, 3, 5, 5, 4, 4$$

are all rounded to the nearest whole number. In this case, the assumption that the measurements represent a random sample from a continuous-valued normal distribution is inappropriate. However, it may be reasonable to assume that the *actual* unrounded UV measurements $x_1, ..., x_{13}$ are independent from a normal distribution with mean μ and variance σ^2, and the recorded measurements are rounded versions of the x_i.

a. Assume that (μ, σ^2) is assigned a noninformative prior proportional to $1/\sigma^2$. Note that if the actual measurements are $N(\mu, \sigma^2)$, the probability of observing the rounded measurement of 7 is equal to

$$\Pr(y_1 = 7|\mu, \sigma^2) = \Pr(6.5 < x_1 < 7.5|\mu, \sigma^2)$$

(2.15)

$$= \Phi\left(\frac{7.5 - \mu}{\sigma}\right) - \Phi\left(\frac{6.5 - \mu}{\sigma}\right),$$

where Φ is the standard normal cdf. Use this expression and the assumption of independent measurements to find the posterior density on (μ, σ^2).

b. Suppose that the unknown unrounded measurements $x_1, ..., x_{13}$ are included into the estimation problem. Suppose that the values of $x_1, ..., x_{13}$ are known. Find the posterior density of (μ, σ^2) conditional on the values of x_i.

c. Suppose, on the other hand, that the parameters (μ, σ^2) are known. Find the posterior density of the unrounded measurements. (These will turn out to be a series of truncated normal distributions.)

d. From parts (b) and (c), outline the construction of a Gibbs sampling algorithm for simulating from the joint distribution of $x_1, ..., x_{13}$ and (μ, σ^2).

3

Regression Models for Binary Data

Suppose that a student is chosen at random from a statistics class. At the end of the course, what is the probability that the selected student passed?

If we knew the final grades of all other students taking the class, a simple way to estimate this probability might be to compute the proportion of all other students who passed and to use that proportion as our estimate. But suppose we wanted to obtain a more accurate estimate. What other sources of information might help us refine our estimate of the probability that the selected student passed, and how could we formally incorporate such information into a statistical model for this probability?

To be more specific, let us suppose we knew the grades of students in a prerequisite probability class, along with their math SAT (SAT-M) scores. Could we use this information to better predict the success of the selected student?

This type of problem can be recast in the following more general way. We have some number n of experimental subjects or observational units. At the end of the experiment or observational period, each of these units is labeled as either a success or failure. In addition, we also observe characteristics of the subjects which may be useful in predicting or explaining their success or failure. In the example above, the units correspond to students in the statistics class, and success is defined by a "passing" grade (say a C or higher). Characteristics relevant to predicting success or failure are the students' grades in a prerequisite probability course and their SAT-M scores.

Data in which each outcome can be described as either a success or a failure are called binary data. Following the notation of previous chapters, we let y_i denote the outcome of the ith experimental or observational unit, and assign to y_i a value of 1 if the ith observation is a success, and 0 if it is a failure. The probability that the ith observation results in a success is denoted by p_i, that is,

$$\Pr(y_i = 1) = p_i.$$

The characteristics that are used to predict the outcome of the binary variables y_i are referred to as *explanatory variables*, or *covariates*. The covariate vector associated with the ith subject is denoted by x_i. In the example above, the covariate vector for the ith individual, x_i, would consist of the individual's grade in the probability class and his or her SAT-M score. Possible values of the covariate vector for two students are (B, 620) and (C, 530).

Binary data are common in designed experiments and observational studies. Besides this somewhat contrived example, in this chapter we will also examine data involving socialization of children in a longitudinal study of inner-city youths. However, in order to avoid the many complications that arise when analyzing real data, we begin our study by examining the stylized problem of predicting the probability that a student receives a grade of C or higher in a statistics class. The data for this example are based on grades assigned to students in a two-semester probability and statistics sequence taught at Duke University. However, to maintain confidentiality of student records, both the SAT-M scores and the grades received by the students in the two-semester sequence have been altered in a way that does not affect the basic conclusions of the analysis.

3.1   Basic modeling considerations

To illustrate the fundamental issues that arise in modeling binary data, consider the grade data displayed in Table 3.1. These data represent slightly disguised versions of student grades in a statistics class, along with transformed versions of their SAT-M scores and the grade received in a prerequisite probability course.

A plot of the passing indicator (i.e., C or higher) variable y_i versus the SAT-M score is displayed in the top panel of Figure 3.1. This plot is difficult to interpret since y_i only takes on two possible values, 0 and 1. To make the relationship between SAT-M score and passing grade clearer, the students were next grouped according to SAT-M score, and the observed proportion of students passing by group was plotted against SAT-M in the bottom panel of the figure.

From Figure 3.1, there appears to be a relationship between SAT-M score and the probability of passing the statistics course. To model this relationship, we must address the following questions:

1. What functional form best describes the relationship between the success probabilities and explanatory variables? That is, how should we *link* the probability of success to the explanatory variables?
2. Given an appropriate functional form describing this relationship, how can we estimate the parameters in the model and how can we assess the uncertainty in our estimates of these parameters?

Student #	Grade	y_i	SAT-M score	Grade in prerequisite probability course
1	D	0	525	B
2	D	0	533	C
3	B	1	545	B
4	D	0	582	A
5	C	1	581	C
6	B	1	576	D
7	C	1	572	B
8	A	1	609	A
9	C	1	559	C
10	C	1	543	D
11	B	1	576	B
12	B	1	525	A
13	C	1	574	F
14	C	1	582	D
15	B	1	574	C
16	D	0	471	B
17	B	1	595	B
18	D	0	557	C
19	F	0	557	A
20	B	1	584	A
21	A	1	599	B
22	D	0	517	C
23	A	1	649	A
24	B	1	584	C
25	F	0	463	D
26	C	1	591	B
27	D	0	.488	C
28	B	1	563	B
29	D	1	553	B
30	A	1	549	A

TABLE 3.1. Hypothetical grades for a class of statistics students. The first column provides an arbitrary student numbering. The second column lists the grade received in the class, and the third column, y_i, indicates if it is passing or failing. The fourth and fifth columns provide the scores on the SAT-Math test and grades for a prerequisite probability course taken by all of the students.

3. How well does the model describe the functional relationship between the success probabilities and the explanatory variables and how well does it predict the variation of individual subjects from their expected outcomes?

3.1.1 Link functions

As an initial step in analyzing binary data, we first focus on the relationship between the success probabilities and explanatory variables, or in the present case, the probability that a student receives a passing grade and the student's SAT-M score.

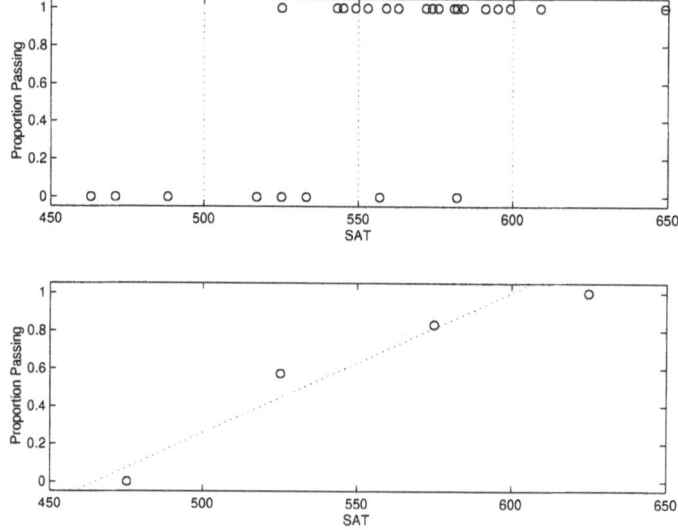

FIGURE 3.1. The top graph plots the indicator of success y_i against SAT-M score. In the bottom panel, students were grouped according SAT-M score, with the groups indicated by the dotted lines in the top plot. The bottom graph displays the proportion of students passing in each group against the midpoint of the grouping interval. The least squares line is depicted by a dotted line in the graph.

Perhaps the simplest assumption that we could make about the relationship between the success probabilities and SAT-M scores would be to assume that the relationship was linear; that is, we might assume that the success probability for the ith student, p_i, could be expressed in the form

$$p_i = \beta_0 + \beta_1 \times \text{SAT-M}_i . \tag{3.1}$$

In this equation, β_0 and β_1 represent unknown constants to be estimated from the data and SAT-M$_i$ denotes the SAT-M score for the ith student. In specifying Equation (3.1), we have assumed that the probability that student i receives a passing grade, p_i, increases (or decreases) linearly with the student's SAT-M score. The quantities β_0 and β_1 are called regression coefficients and specify the theoretical, but unknown, equation of the line that best describes this linear relationship. Referring again to Figure 3.1, the assumption that there is a linear relationship between the success probabilities p_i and SAT-M scores appears, at first glance, to be quite reasonable.

Figure 3.1 depicts the least squares line through this data. The equation for this line is

$$\hat{p}_i = -3.44 + 0.0074 \times \text{SAT-M}_i. \tag{3.2}$$

From this equation, we may predict the probability that a future student with a SAT-M score of 580 receives a passing grade is

$$\hat{p}_i = -3.44 + 0.0074 \times 580 = .85. \tag{3.3}$$

Thus, it appears that such a student is quite likely to pass the course.

But what of a student who receives a 350 SAT-M score? According to Equation (3.2), this student passes the course with probability $-3.44 + .0074 \times 350 = -.85$. Similarly, a student who receives an 800 SAT-M score is estimated to have probability $-3.44 + .0074 \times 800 = 2.48$ of passing.

Obviously, we would like probabilities predicted by the regression model to lie within the interval $(0, 1)$. The fact that our linear regression equation predicts values outside of this interval suggests that relationships between explanatory variables and success probabilities cannot, in general, be modeled using a straight line.

One solution to this problem is to model success probabilities using functions that take values only in the interval $(0, 1)$. For statisticians, a natural class of such functions is provided by the class of cumulative distribution functions, or *cdf's*. Recall that for any random variable X, the cumulative distribution function of X at a point a, denoted $F(a)$, is defined to be the probability that X is less than or equal to a; that is, $F(a)$ is defined as $F(a) = \Pr(X \le a)$.

To incorporate a cdf into the relationship between the success probabilities and explanatory variables, we modify (3.1) to reflect the behavior of F by assuming that

$$p_i = F(\beta_0 + \beta_1 \times \text{SAT-M}_i). \tag{3.4}$$

In (3.4), F is called the *link function* because it links the linear function of the covariates to the success probability.

There are, of course, a large number of cdf's that might be used to model success probabilities, and except for certain retrospective study designs, there are seldom substantive grounds to prefer one over another. Thus, the choice of link function is often arbitrary. For this reason, it is standard practice to fit data using several link functions and to choose the link function that yields the best fit of the model to the data. Not surprisingly, a common choice for the link function F is the standard normal cdf, which we denote by Φ. Binary regression models that employ this link function are called *probit models*.

To illustrate how the probit link function can be used to model success probabilities, model (3.4) was applied to the grade data of Table 3.1 with $F(\cdot) = \Phi(\cdot)$. Maximum likelihood procedures were used to estimate the regression coefficients β_0 and β_1, which, in this case, turn out to be -17.96 and 0.0334. Thus, the maximum likelihood estimates of the success probabilities p_i are

$$\hat{p}_i = \Phi(-17.96 + 0.0334 \times \text{SAT-M}_i). \tag{3.5}$$

Substantively, this equation may be interpreted in one of several ways. For instance, using (3.5), a student who scores 600 on her SAT-M will pass the course with an estimated probability of $\Phi(-17.96 + 0.0334 \times 600) = .98$. Students who score 350 and 800 on the SAT-M are similarly predicted to pass with probabilities less than 0.001 (but greater than 0) and greater than .999 (but less than 1), respectively. The slope parameter β_1 determines the number of SAT-M points required to increase the probability of passing by one unit on the standard normal scale; that is, for every $1/\beta_1 \approx 30$ unit increase in math SAT, a student's pass probability increases by one unit on the standard normal scale.

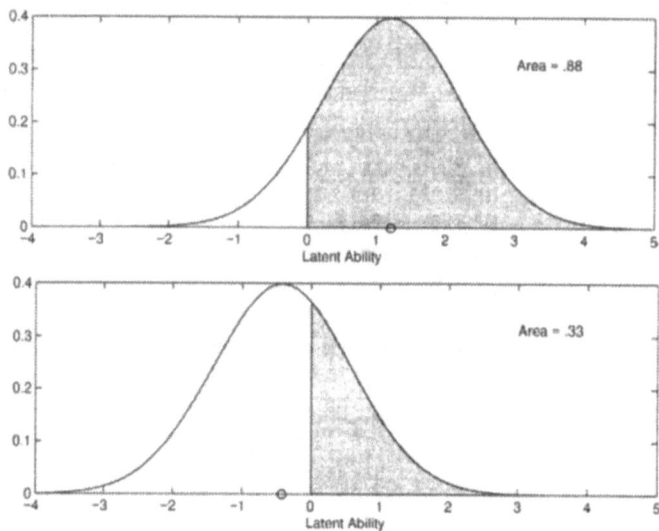

FIGURE 3.2. Illustration of probit success probabilities for two students. The normal densities represent the distribution of the latent abilities and the shaded areas represent the probabilities that the students pass.

A geometrical interpretation of probit regression is provided in Figure 3.2. This figure illustrates the pass probabilities for two students who scored 574 and 525 on their math SAT tests. The linear predictors of the success probabilities for the two students, on the standard normal scale, are $-17.96 + 0.0334 \times 574 = 1.21$ and $-17.96 + 0.0334 \times 525 = -.43$. These values may be interpreted as latent (unobserved) ability variables that underlie the performance of these two students in the statistics class. In Figure 3.2, the normal densities centered at 1.21 and $-.43$ depict the distribution of abilities for students with the corresponding SAT-M scores. If the latent value drawn for a student falls above zero, the student is assumed to pass the course. This event happens with probability equal to the shaded area in the plots. If the latent value drawn for an individual falls below zero, a failure occurs. By comparing the two graphs, we see that the latent performance distribution for students with higher SAT-M scores have greater mass above zero, and, therefore, these students are predicted to have a higher probability of passing. In general, as SAT-M score increases, the center of the normal performance density moves to the right at rate β units per SAT point, and the associated pass probability increases accordingly.

Besides the probit link, another commonly used link function is the logistic link function. The logistic link function is derived from the standard logistic distribution function, which can be expressed as

$$F(x) = \frac{e^x}{1 + e^x}, \qquad -\infty < x < \infty. \tag{3.6}$$

The simple form of this distribution allows us to express the success probabilities explicitly in terms of the linear predictor. For the grade data, this expression is

$$\log\left(\frac{p_i}{1-p_i}\right) = \beta_0 + \beta_1 \times \text{SAT-M}_i. \tag{3.7}$$

The quantity $\log(p_i/(1-p_i))$ is called the *logistic transformation* of the success probability p_i, or the *logit* for short.

Another commonly used link function is based on the extreme-value distribution

$$F(x) = 1 - \exp(-\exp(x)), \qquad -\infty < x < \infty. \tag{3.8}$$

This distribution function leads to a model for the success probabilities that has the form

$$\log[-\log(1-p_i)] = \beta_0 + \beta_1 \times \text{SAT-M}_i. \tag{3.9}$$

This link is often called the complementary log-log link function. Unlike the normal and logistic distributions, the extreme-value distribution is asymmetric around the value 0.

How does the choice of link function affect the fit of the regression model? To explore this question, the three link functions described above were each fit to the statistics class data using maximum likelihood estimation. The fit of the probit model was described above. The fitted model equations for the logit and complementary log-log links are

$$\log\left(\frac{p_i}{1-p_i}\right) = -31.115 + 0.0578 \times \text{SAT-M}_i,$$

$$\log[-\log(1-p_i)] = -17.836 + 0.0323 \times \text{SAT-M}_i.$$

At first glance, the disparate values of these regression parameters appear to indicate large differences between the fitted models. However, the parameters cannot be compared directly because of differences in scaling between the standard forms of the underlying distribution functions. For example, the standard deviation of the standard logistic distribution function is $\pi/\sqrt{3}$, while the standard deviation of the extreme value distribution is $\pi/\sqrt{6}$, and the difference between the logistic and extreme value regression estimates is primarily a reflection of these different spreads. Thus, comparisons between the three models are more appropriately made using predicted probabilities. The fitted values for the success probabilities under these three models are displayed in Figure 3.3. It is noteworthy that predictions from the probit and logit models are barely distinguishable for moderate values of SAT-M and differ significantly only in their relative values for predicted probabilities close to 0 or 1. In contrast, the complementary log-log link gives substantively different fits than the probit or logit. Its fitted probabilities are larger for extreme values of SAT-M scores and smaller for SAT-M scores in the range of 540–580.

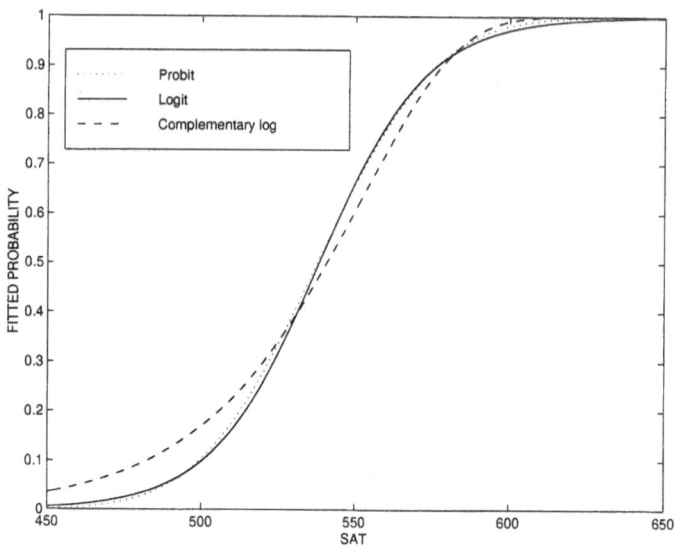

FIGURE 3.3. Graph of fitted probabilities using the logit, probit, and complementary log-log models for the statistics class dataset.

3.1.2 Grouped data

In many studies, individuals or experimental units may be grouped according to observed covariates. For example, in predicting the probability that students receive passing grades based on their SAT-M scores, it may happen that several students have the same SAT-M score. In the binary regression model, such students would have the same predicted probability of passing.

The presence of experimental units that can be grouped according to common covariate values has important implications in assessing model fit. For this reason, it is useful to expand our notation slightly to accommodate grouped binary data, or binomial data.

To this end, we now modify the definition of y_i to denote the number of successes observed in the ith experimental group. If there is a single individual in the group, y_i has the same meaning as before. The number of subjects in the ith group is denoted n_i; for groups of size one, $n_i = 1$. The total number of groups is denoted by n, the total number of observations by M. Note that $M = \sum_{i=1}^{n} n_i$.

3.2 Estimating binary regression coefficients

3.2.1 The likelihood function

In Section 3.1.1, we found that the relationship between the explanatory variables in a binary regression model and the success probabilities could not be adequately modeled through a simple linear relationship. For similar reasons, the least squares criteria for estimating regression coefficients in classical regression settings is not appropriate for binary regression models.

In standard least squares regression models, observational errors are assumed to have a Gaussian distribution, and regression coefficients are estimated using least squares. Continuing the example of the last section, suppose we modeled student GPA as a linear function of combined SAT-M score and random error through the relation

$$\text{GPA}_i = \beta_0 + \beta_1 \text{SAT-M}_i + \epsilon_i, \qquad i = 1, \ldots, n. \tag{3.10}$$

We might also assume that the distribution of the error term, ϵ_i, is a Gaussian distribution with mean 0 and unknown variance σ^2. Since the density of a $N(0, \sigma^2)$ random variable is

$$f(x) = \frac{1}{\sqrt{2\pi}\sigma} \exp\left(\frac{-x^2}{2\sigma^2}\right),$$

by solving for ϵ_i in (3.10), we find that the likelihood function for β_0, β_1, and σ^2 is

$$\prod_{i=1}^{n} \frac{1}{\sqrt{2\pi}\sigma} \exp\left[-\frac{(\text{GPA}_i - \beta_0 - \beta_1 \times \text{SAT-M}_i)^2}{2\sigma^2}\right]. \tag{3.11}$$

Taking logarithms, it follows that the log-likelihood function is

$$l(\beta_0, \beta_1, \sigma^2) = n\log(\sigma) - \frac{1}{2\sigma^2}\sum_{i=1}^{n}(\text{GPA}_i - \beta_0 - \beta_1 \times \text{SAT-M}_i)^2. \tag{3.12}$$

The maximum likelihood estimates of the parameters β_0, β_1, and σ^2 are the values of the parameters that maximize the likelihood function or, equivalently, that maximize the log-likelihood function. From (3.12), we find that the log-likelihood function, for any fixed value of σ^2, is maximized when

$$\sum_{i=1}^{n}(\text{GPA}_i - \beta_0 - \beta_1 \times \text{SAT-M}_i)^2 \tag{3.13}$$

is minimized as a function of β_0 and β_1. The values of β_0 and β_1 that minimize the sum of squares (3.13) are thus the maximum likelihood estimates and are also the least squares estimates; that is, in simple linear regression models with independent, identically distributed normal errors, the least squares estimates and the maximum likelihood estimates are equivalent.

In the case of binary regression models, however, the least squares estimates do not correspond to maximum likelihood estimates, or any other likelihood-based procedure.

To derive the likelihood function for binary regression models, recall that for a binomial random variable y, which denotes the number of successes in N trials each having success probability p, the sampling density can be expressed as

$$f(y|p) = \binom{N}{y} p^y (1-p)^{N-y}, \qquad y = 0, \ldots, N. \tag{3.14}$$

The likelihood function is proportional to the sampling density but is viewed as a function of p:

$$L(p) \propto p^y (1-p)^{N-y}, \ 0 < p < 1.$$

In the regression setting, we assume that success probabilities vary from observation to observation according to the relationship

$$p_i = F(\mathbf{x}_i'\beta), \tag{3.15}$$

where $F(\cdot)$ denotes the link function and $\mathbf{x}_i'\beta$ is shorthand notation for the linear combination of explanatory variables. For predicting the probabilities of students passing a statistics course based on their SAT-M scores,

$$\mathbf{x}_i'\beta = \beta_0 + \beta_1 \times \text{SAT-M}_i.$$

In more general problems containing r explanatory variables,

$$\mathbf{x}_i'\beta = \beta_0 + \beta_1 x_{i1} + \beta_2 x_{i2} + \cdots + \beta_r x_{ir},$$

where $x_{i1}, ..., x_{ir}$ are the values of the covariates corresponding to the ith individual.

For n independent observations with success probabilities described by (3.15), the likelihood function for β, given $\mathbf{y} = \{y_1, y_2, \ldots, y_n\}$, can be obtained by substituting (3.15) into (3.14) and multiplying over the observations:

$$L(\beta) = \prod_{i=1}^{n} F(\mathbf{x}_i'\beta)^{y_i} (1 - F(\mathbf{x}_i'\beta))^{n_i - y_i}. \tag{3.16}$$

Based on the likelihood function (3.16), parameter estimation and inference can proceed in one of two ways—either inference can be based on maximum likelihood estimation or it can be based on Bayesian methodology. For simple binary regression models, substantitive conclusions from the two analyses can be quite similar; although from a philosophical perspective, the interpretation of these conclusions may be quite different. In either case, maximum likelihood estimates provide a convenient starting point for many of the numerical procedures used to implement a fully Bayesian analysis and generally provide an accurate approximation to Bayesian posterior densities when vague or noninformative priors are used.

3.2.2   Maximum likelihood estimation

Historically, interest in binary regression models and other *generalized linear models* intensified after the discovery that maximum likelihood estimates could be obtained using modified least squares procedures (Nelder and Wedderburn, 1972). Based on Newton-Raphson function maximization, these iteratively reweighted least squares routines offered statisticians an opportunity to routinely fit binary regression models at a time when computational resources were scarce.

The appendix to this chapter outlines an iterative algorithm that can be used to find the maximum likelihood estimate $\hat{\beta}$ of the regression parameter vector β. As a byproduct of this fitting algorithm, one obtains the asymptotic covariance matrix $\hat{C}$. Inference about the regression vector can based on the asymptotic normality of the sampling distribution of the MLE $\hat{\beta}$. Specifically, $\hat{\beta}$ has an approximate normal sampling distribution with mean β and covariance matrix $\hat{C}$. This distribution is also useful in finding interval estimates for components of β and in performing significance tests concerning the value of β.

3.2.3   Bayesian estimation and inference

Inference based on maximum likelihood estimation requires the specification of the conditional distribution of the data, given the values of the model parameters. Bayesian inference depends on the joint specification of both the data and the model parameters. The additional "price" of Bayesian inference is, thus, the requirement to specify the marginal distribution of the parameter values, or the prior. The return on this investment is substantial. We are no longer obliged to rely on asymptotic arguments when performing inferences on the model parameters, but instead can base these inferences on the exact conditional distribution of the model parameters given observed data—the posterior.

Prior distributions

To perform Bayesian inference in the binary regression setting, we must add to our model formulation the specification of a prior density on the regression parameter vector β. If approached directly, this prior specification can be a difficult task due to the indirect effect that the regression parameters have on the success probabilities. However, this difficulty can be circumvented by use of the *conditional means family* of prior densities (see, for example, Bedrick, Christensen, and Johnson, 1996). In this class of distributions, prior beliefs about the location of the success probabilities p_i are assessed for particular values of the covariates x_i, and this information is used to construct a prior for the regression parameter vector β. We illustrate the use of this prior formulation here since it is one of the easiest methods of incorporating subjective prior knowledge into the binary regression problem. Of course, a less subjective approach to Bayesian inference can always be executed by simply assuming a uniform prior on the regression vector β, as one implicitly does when computing the maximum likelihood estimate.

Suppose, then, that there are r explanatory variables $x_1, ..., x_r$. To subjectively input prior information, we consider $b = r + 1$ different values of the covariate vector. In our statistics class example, there is $r = 1$ explanatory variable, SAT-M, and so we examine $b = 2$ different values of the SAT-M score. For each value of the covariate vector, we specify two values: (i) a guess at the probability of success p_i—call this guess g_i and (ii) a statement about how sure we are of this guess. We state this sureness in terms of the number of equivalent "prior observations." Denote this prior sample size by K_i.

In the statistics class example, suppose that we chose the two SAT-M values 500 and 600 as the values for which we were willing to assess prior probabilities of success. For each of these SAT-M scores, we then estimate the probability that a student with these scores would pass the class. In addition, we indicate how many observations our estimates are worth. For example, we might estimate that a student with a 500 SAT-M score had probability 0.3 of passing the course, whereas a student with a score of 600 might have a pass probability of 0.7. If we were fairly confident in these estimates, we might value them as the equivalent of, say, five observations. This prior input corresponds to values $g_1 = .3$, $g_2 = .7$, $K_1 = 5$, and $K_2 = 5$.

To transform these guesses into a formal prior density on β, we match this prior information to beta densities with parameters $K_i g_i$ and $K_i(1 - g_i)$. If we assume that the b probabilities $p_1, ..., p_b$ are a priori independent, the joint density of the probabilities is given by

$$g(p_1, ..., p_b) \propto \prod_{i=1}^{b} p_i^{K_i g_i - 1}(1 - p_i)^{K_i(1-g_i)-1}. \tag{3.17}$$

This distribution on the probabilities $\{p_i\}$ implies a prior on the regression vector β. Letting $F(\cdot)$ denote the link distribution function and $f(\cdot)$ its derivative, the induced prior based on (3.17) is

$$g(\beta) \propto \prod_{i=1}^{b} F(\mathbf{x}_i'\beta)^{K_i g_i - 1}(1 - F(\mathbf{x}_i'\beta))^{K_i(1-g_i)-1} f(\mathbf{x}_i'\beta). \tag{3.18}$$

This particular choice of prior is based on the assumption that some prior information about the location of the probabilities of success for selected values of the covariates is available, even if the associated precision of this prior information is small (i.e., $K_i < 1$). However, if no prior information is available, then the usual vague prior for the regression parameter β can be used as a default by taking

$$g(\beta) = 1.$$

The posterior distribution

Given a prior density $g(\beta)$, the posterior density is proportional to

$$g(\beta|\text{data}) \propto L(\beta)g(\beta), \tag{3.19}$$

where the likelihood function $L(\beta)$ is specified in (3.16) and "data" refers to the binomial observations $y_1, ..., y_n$. If a uniform prior is specified, then the posterior density is proportional to the likelihood function. As in the case of maximum likelihood estimates, the posterior density for the regression parameter in binary models cannot be derived analytically. Instead, numerical techniques are needed to summarize this probability distribution.

Simulating from the posterior distribution

A variety of Gibbs sampling and Metropolis-Hastings algorithms have been proposed for obtaining samples from the joint posterior distribution on regression parameters in binary models. Here, we consider a simple variation of the

Metropolis-Hastings sampler which appears to work well for binary regression problems where the number of covariates does not exceed 10.

This Metropolis-Hastings simulation algorithm uses the maximum likelihood estimate $\hat{\beta}$ as its starting value and the asymptotic covariance matrix $\hat{C}$ of the maximum likelihood estimate in the specification of the proposal density. In the following, let σ_{MH} denote an adjustable parameter in the Metropolis-Hastings scheme that controls the width of the proposal density, and suppose that m updates are required. With this notation, the generic random-walk Metropolis-Hastings algorithm for sampling from the posterior on the regression parameters may be specified by the following steps.

Metropolis-Hastings sampler:

0. Set $\beta^{(0)} = \hat{\beta}$, $j = 0$, $a = 0$, and $\sigma_{MH} = 1.0$.
1. Increment $j = j + 1$.
2. Draw $\beta^c \sim N(\beta^{(j-1)}, \sigma_{MH}\hat{C})$ and $u \sim U(0, 1)$.
3. Calculate

$$r = \frac{L(\beta^c)g(\beta^c)}{L(\beta^{(j-1)})g(\beta^{(j-1)})}. \tag{3.20}$$

4. If $u < r$, set $\beta^{(j)} = \beta^c$. Otherwise, set $\beta^{(j)} = \beta^{(j-1)}$ and $a = a + 1$.
5. If $j < m$, return to (1).

Steps (1)–(4) should be repeated a large number of times, until an adequate sample size is obtained. In determining this sample size, the number of accepted draws should be monitored and, if necessary, the value of σ_{MH} should be adjusted so that the acceptance rate a/m falls in the range $(0.25, 0.50)$. Decreasing σ_{MH} increases the acceptance ratio, and increasing σ_{MH} decreases it. For low-dimensional parameter vectors, values of σ_{MH} in the range $(1, 2)$ often produce acceptable results, but as the dimension of the parameter vector increases, the value of σ_{MH} must usually be decreased.

3.2.4 An example

We now apply these estimation techniques to the pass/fail data from the statistics class. A logistic model of the form

$$\log\left(\frac{p_i}{1 - p_i}\right) = \beta_0 + \beta_1 \times \text{SAT-M}_i \tag{3.21}$$

is assumed. The maximum likelihood estimates of the intercept β_0 and the slope β_1 are given by -31.11 and $.0578$, respectively. The asymptotic standard errors of these estimates (obtained from the estimated covariance matrix $\hat{C}$) are 12.56 and .0255, respectively.

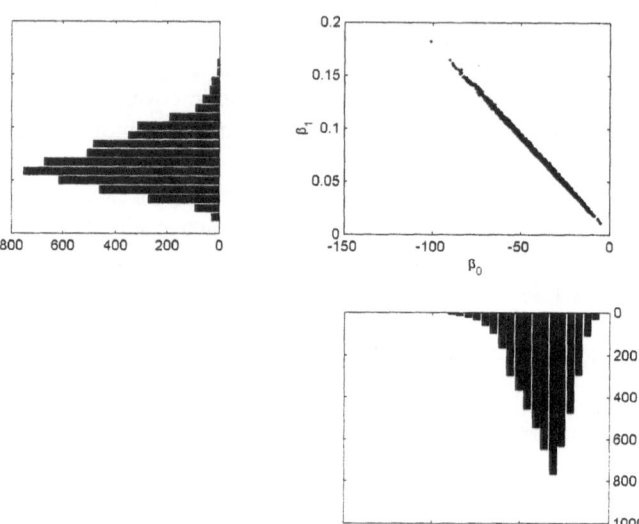

FIGURE 3.4. Scatterplot and marginal histograms of the simulated values of the posterior distribution of (β_0, β_1) using a uniform prior.

We consider two Bayesian analyses: one using a uniform prior and one using an informative prior. Consider first the uniform prior. Applying the Metropolis-Hastings algorithm initialized at the MLE, we generated 5000 sample points, denoted $\{(\beta_0^{(j)}, \beta_1^{(j)}), j = 1, ..., 5000\}$. Ideally, these points represent 5000 correlated draws from the joint posterior distribution on (β_0, β_1). Figure 3.4 depicts the values obtained through this procedure.

All of the simulated parameter values lie near a line with negative slope, which indicates that the intercept and slope parameters are negatively correlated. This figure also depicts histograms of the simulated values of β_0 and β_1—these histograms are pictures of the marginal posterior densities of the two parameters. Note that both marginal densities are skewed away from the origin, suggesting that both the normal approximation to the posterior density and the asymptotic normal distribution of the MLE are unlikely to be very accurate.

Table 3.2 displays summaries of the marginal posterior distributions for the intercept and slope parameters. The posterior mean, standard deviations, and selected quantiles are provided. Note that the posterior mean of β_1 is .0695, whereas the posterior mode (which is equivalent to the MLE) is somewhat smaller, taking values of .0578. The quantiles provided in the table can be used for interval estimation. For example, a 95% probability interval for the slope is (.0282, .1266).

The simulated sample from the posterior distribution can also be used to obtain samples from any function of the parameters of interest. To illustrate, suppose we are interested in estimating the fitted probabilities p_i for all students. These

Method		β_0 (slope)	β_1 (intercept)
Maximum likelihood estimate (se)		−31.11(12.56)	.0578 (.0228)
Bayes (flat prior)	Mean (st. dev.)	−37.45(14.04)	.0695 (.0255)
	2.5 percentile	−68.36	0.0282
	5 percentile	−61.99	0.0326
	50 percentile	−35.67	0.0661
	95 percentile	−17.26	0.1142
	97.5 percentile	−14.73	0.1266

TABLE 3.2. Maximum likelihood and Bayesian summary statistics for fits of logit model to statistics class dataset.

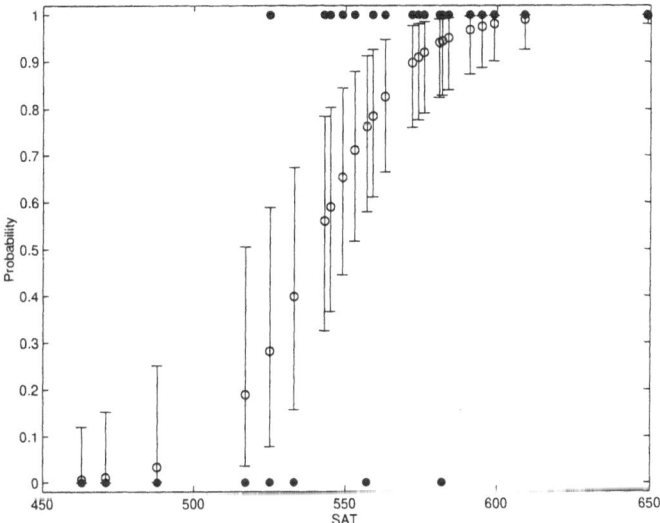

FIGURE 3.5. Line plots of the posterior distributions for the fitted probabilities p_i. Each line plot shows the location of the 5th, 50th, and 95th percentiles of the distribution. The observed values of y_i are shown as solid dots.

probabilities are given by

$$p_i = \frac{\exp(\beta_0 + \beta_1 \times \text{SAT-M}_i)}{1 + \exp(\beta_0 + \beta_1 \times \text{SAT-M}_i)}. \tag{3.22}$$

By substituting the simulated values of β_0 and β_1 into (3.22), we obtain histogram estimates of the posterior distribution for the fitted probability of success for each student. These distributions are displayed in Figure 3.5. Each posterior is depicted as a vertical line; the median of each distribution is indicated by a circle, and the endpoints of the line segments correspond to the 5th and 95th percentiles of the distributions. Note the skewness of the posterior densities concentrated near 0 and 1.

A Bayesian analysis using this informative prior density proceeds along similar lines. As discussed above, we assume that our prior beliefs about the effect of SAT-M on passing the statistics course can be summarized through the assumptions that a student with a 500 SAT-M score will pass the course with probability .3, that a student with a 600 SAT-M score passes with probability .7, and each probability estimate is "worth" about 5 observations. From these prior opinions, we constructed a prior density on β using Equation (3.17) and then reran the Metropolis-Hastings algorithm using the revised posterior density. Figure 3.6 summarizes the results of the informative and noninformative Bayesian models from 5000 samples from this Metropolis-Hastings. The two P's in the graph represent the two prior probability estimates. The solid line in this graph represents the posterior means of the fitted probabilities using the uniform prior. The posterior means of β_0 and β_1 using the informative prior are -20.46 and $.0383$, respectively, and the corresponding posterior means of the fitted probabilities are also shown on the graph using a dotted line. Observe that the posterior fitted probabilities using the informative prior represent a compromise between the posterior probabilities obtained under a uniform prior and the prior probability estimates.

3.3   Latent variable interpretation of binary regression

A concept that is useful in motivating the link functions described in Section 3.1.1 and that plays a prominent role in modeling the more complicated ordinal

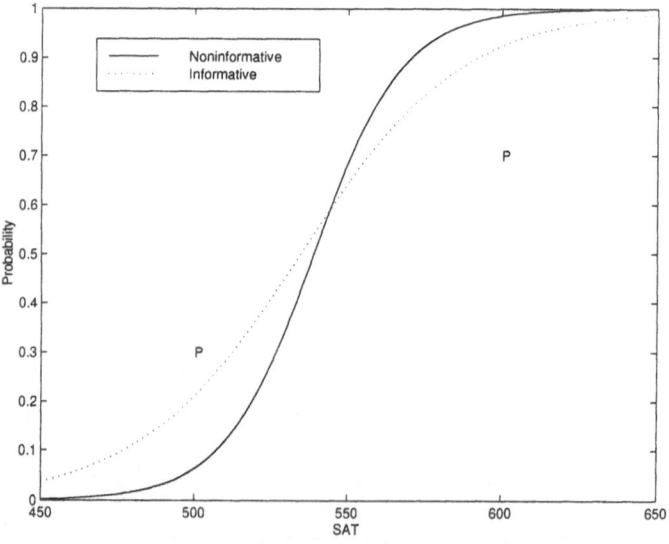

FIGURE 3.6. Fitted probabilities using Bayesian analyses using flat and informative prior distributions. The two prior probability estimates are shown with a P.

data structures described later in this text involves the notion of an underlying latent variable. Latent variables, or latent traits, are used to model unobserved characteristics of subjects that determine corresponding probabilities of success. For models with more than two ordinal response categories, latent variables extend naturally to describe probabilities that observations fall into each category.

Referring again to Figure 3.2, suppose that the success probability for a binary variable y_i is modeled as

$$p_i = F(\mathbf{x}_i'\beta),$$

where $\mathbf{x}_i$ is the vector of explanatory variables, β is a binary regression parameter, and F is a known distribution. To introduce a latent variable structure within this model, for each observation i we introduce a continuous variable Z_i, where

$$Z_i = \mathbf{x}_i'\beta + \epsilon_i, \tag{3.23}$$

where $\epsilon_1, ..., \epsilon_n$ are assumed to be independent and identically distributed according to F. We then assume that

$$y_i = \begin{cases} 0 & \text{if } Z_i \leq 0 \\ 1 & \text{if } Z_i > 0. \end{cases} \tag{3.24}$$

Thus, $Z_i > 0$ if and only if the ith observation is a success. It follows that

$$p_i = \Pr(y_i = 1) = \Pr(Z_i > 0) = 1 - F(-\mathbf{x}_i'\beta).$$

For distributions F that are symmetric about 0, this is equivalent to $p_i = F(\mathbf{x}_i'\beta)$. Thus, the latent variable structure yields a model equivalent to a logistic or probit model, and with only minor and obvious redefinition of the regression parameters, can also be used to motivate asymmetric links like the complementary log-log.

Albert and Chib (1993) discuss the utility of latent variable models for binary and polychotomous data both from a computational viewpoint and as diagnostic tools for model assessment.

To demonstrate the utility of the latent variable approach to modeling binary data, let us confine our attention for the moment to the probit model so that $F = \Phi$, and assume a uniform prior is taken for β. In this case, a Gibbs sampling scheme can be used to simulate from the joint distribution of the vector of latent traits $\mathbf{Z}$ and β by successively simulating from the posterior distribution of $\mathbf{Z}$ given β, and β given $\mathbf{Z}$. The conditional distribution of β given $\mathbf{Z}$ is

$$\beta \sim N(\beta_{LS}, (\mathbf{X}'\mathbf{X})^{-1}), \tag{3.25}$$

where β_{LS} is the least squares estimate of β based on $\mathbf{Z}$, i.e., $(\mathbf{X}'\mathbf{X})^{-1}\mathbf{X}'\mathbf{Z}$ and $\mathbf{X}$ is the covariate matrix with rows $\mathbf{x}_1', ..., \mathbf{x}_n'$.

Similarly, by examining Figure 3.2, one finds that the conditional distribution of each component of $\mathbf{Z}$, Z_i, given y_i and β, has a truncated normal distribution with mean $\mathbf{x}_i'\beta$. Truncation is to the left by 0 if $y_i = 1$, and to the right of 0 if

$y_i = 0$. Equivalently, the conditional density of Z_i given β is proportional to

$$Z_i \mid \beta, y_i \propto \begin{cases} \phi(z; \mathbf{x}_i'\beta, 1)I(z \leq 0) & \text{if } y_i = 0 \\ \phi(z; \mathbf{x}_i'\beta, 1)I(z > 0) & \text{if } y_i = 1, \end{cases} \qquad (3.26)$$

where $\phi(\cdot; a, b)$ denotes a normal density with mean a and variance b and $I(\cdot)$ is the indicator function.

These remarks motivate a Gibbs sampling algorithm for simulating m iterates from the posterior density resulting from a probit model under a uniform prior.

Gibbs sampler for a probit model:

0. Set $\beta^{(0)} = \hat{\beta}$ and $j = 0$.
1. Increment $j = j + 1$.
2. Simulate independent latent variables $Z_1^{(j)}, ..., Z_n^{(j)}$, where $Z_i^{(j)}$ is distributed $\phi(z; \mathbf{x}_i'\beta^{(j-1)}, 1)I(z \leq 0)$ if the corresponding y_i is a failure ($y_i = 0$), and Z_i is distributed $\phi(z; \mathbf{x}_i'\beta^{(j-1)}, 1)I(z > 0)$ if $y_i = 1$.
3. Simulate $\beta^{(j)} \sim N(\beta_{LS}^{(j)}, (\mathbf{X}'\mathbf{X})^{-1})$, where $\beta_{LS}^{(j)} = (\mathbf{X}'\mathbf{X})^{-1}\mathbf{X}'\mathbf{Z}^{(j)}$.
4. If $j < m$, return to (1).

The availability of full-conditional distributions (the multivariate normal distribution on β and truncated normal conditional distribution on the components of Z_i) makes it possible to sample from the posterior distribution on β using Gibbs sampling techniques in lieu of the less efficient Metropolis algorithm described for the logistic model. More importantly, however, the introduction of the latent variables $\mathbf{Z}$ provides us with a useful tool for model diagnostics, because the nominal distribution of the residuals $\mathbf{Z} - \mathbf{X}\beta$ is $N(0, I)$, where I is the identity matrix. This fact allows us to examine standard normal-theory residual diagnostics and plots when assessing model fit, as discussed in Section 3.4.1.

3.4   Residual analysis and goodness of fit

Like the iterative processes described above for parameter estimation, the process of model fitting and evaluation is also an iterative one, in which models are specified, fit to data, and revised.

In general, modeling begins with an exploratory examination of the data, often performed graphically. Following this, a tentative model is specified and fit to the data. Based on the fit, the adequacy of the model is evaluated, and alternative models are devised to address deficiencies in model assumptions. This process continues until either a satisfactory model is obtained or the failures of the final model to explain features of the data are understood and documented.

In simple binary regression, model criticism and selection often focus on assessing the adequacy of a model in predicting the outcome of individual data points, summarizing the fit of a model to the dataset as a whole, and choosing between models containing different subsets of explanatory variables. Examining the adequacy of the model in predicting individual data points falls under the heading of residual or case analysis, while goodness-of-fit statistics are used to summarize overall model adequacy.

3.4.1 Case analysis

Case analyses are used to identify observations, or, in some instances, groups of observations, that deviate from model assumptions. As in the case of linear regression models, unusual observations, or outliers, can affect model assessment in a number of ways: they can unduly affect estimates of regression parameters that would be appropriate for "usual" cases; they can result in inflated estimates of dispersion parameters; or most often, they may signal deficiencies in the model itself. In all cases, identifying outlying observations is important.

Like the techniques used in formulating and fitting binary regression models, residual analyses for binary data have historically drawn on techniques developed for least squares analyses. As a result, the most common approach toward analyzing residuals from a binary regression equation is to transform the binary residuals to a scale upon which they are approximately normally distributed and to then analyze the transformed residuals using techniques employed for normal-theory models. The normalizing transformations most commonly used in this pursuit lead to Pearson, deviance, and adjusted deviance residuals.

Unfortunately, these transformations are useful only in analyzing residuals for binomial observations with sample sizes greater than about five. When the binomial sample size is less than five, and, in particular, for Bernoulli observations, such transformations are hardly "normalizing," and the distributions of Pearson, deviance, and adjusted deviance are then not well approximated by a Gaussian distribution. In such cases, classical residual analyses fail, and it is necessary to turn instead to a fully Bayesian residual analysis. Bayesian residual analyses focus on the posterior distribution of each residual and aim at identifying those residuals whose distribution is concentrated on intervals not containing the point zero.

We now examine each of these residuals in detail.

Pearson residuals

Pearson residuals, or standardized residuals, are obtained by dividing the ith residual, canonically defined as

$$y_i - \hat{y}_i,$$

by its estimated standard error. Assuming a model of the form

$$p_i = F(\mathbf{x}_i'\beta)$$

and letting $\hat{\beta}$ denote the MLE of the regression parameter, the fitted value for the ith binomial observation may be written as

$$\hat{y}_i = n_i \, \hat{p}_i, \qquad \text{where} \qquad \hat{p}_i = F(\mathbf{x}_i'\hat{\beta}).$$

Because the variance of a binomial observation is $n_i p_i (1 - p_i)$, Pearson residuals for binary regression models are defined as

$$r_{i,P} = \frac{y_i - \hat{y}_i}{\sqrt{n_i \hat{p}_i (1 - \hat{p}_i)}}. \tag{3.27}$$

As mentioned above, for $n_i \geq 5$, or, more accurately, when $n_i p_i \geq 3$ and $n_i (1 - p_i) \geq 3$, the distribution of Pearson residuals can be reasonably approximated by a standard normal distribution. In such cases, Pearson residuals with absolute values greater than two or three might be regarded with suspicion.

Deviance residuals

Deviance residuals are related to the deviance function, D, which is defined as twice the difference of the log-likelihood function evaluated at the observed proportions and the model's fitted values; that is,

$$D = 2 \sum_{i=1}^{n} \left\{ y_i \log \left(\frac{y_i}{\hat{y}_i} \right) + (n_i - y_i) \log \left(\frac{n_i - y_i}{n_i - \hat{y}_i} \right) \right\}. \tag{3.28}$$

The deviance residual, $r_{i,D}$, is defined for the ith observation as the signed square root of the ith observation's contribution to total model deviance, or

$$r_{i,D} = \text{sign}(y_i - \hat{y}_i) \left\{ 2 \left[y_i \log \left(\frac{y_i}{\hat{y}_i} \right) + (n_i - y_i) \log \left(\frac{n_i - y_i}{n_i - \hat{y}_i} \right) \right] \right\}^{1/2}, \tag{3.29}$$

where $\text{sign}()$ is 1 (-1) if the argument is positive (negative). Like the Pearson residual, the deviance residual is approximately normally distributed when $n_i p_i \geq 3$ and $n_i (1 - p_i) \geq 3$.

Adjusted deviance residuals

Adjusted deviance residuals (Pierce and Schaffer, 1986) are obtained from deviance residuals by making a first-order correction for the mean bias. For binary regression models, this leads to adjusted deviance residuals of the form

$$r_{i,AD} = r_{i,D} + \frac{1 - 2\hat{p}_i}{6\sqrt{n_i \hat{p}_i (1 - \hat{p}_i)}}. \tag{3.30}$$

Of the three residuals described, the distribution of the adjusted deviance residuals is often most nearly normal and yields surprisingly accurate approximations for even small n_i, provided that the fitted probabilities are not too close to 0 or 1.

Table 3.3 provides a comparison of the three transformed residuals for the maximum likelihood fit (using a logit link) to the data of Table 3.1. A plot of the deviance residuals versus the fitted values appears in Figure 3.7. The pattern of the curves in this figure is typical of residual plots for binary observations. The upper curve in this figure corresponds to residuals of success observations and the lower curve the residuals of the failures. From this plot, four residuals appear extreme—three at the lower right section of the graph and one in the upper left section. It is difficult

Student #	Observed	Fitted	Pearson	Deviance	Adj. Dev.
1	0	0.322	−0.689	−0.882	−0.755
2	0	0.430	−0.869	−1.061	−1.014
3	1	0.602	0.813	1.008	0.938
4	0	0.928	−3.585	−2.293	−2.844
5	1	0.924	0.287	0.398	−0.135
6	1	0.901	0.332	0.457	0.010
7	1	0.878	0.372	0.510	0.124
8	1	0.984	0.128	0.180	−1.104
9	1	0.773	0.542	0.718	0.501
10	1	0.574	0.862	1.054	1.004
11	1	0.901	0.332	0.457	0.010
12	1	0.322	1.450	1.505	1.632
13	1	0.890	0.352	0.483	0.067
14	1	0.928	0.279	0.387	−0.164
15	1	0.890	0.352	0.483	0.067
16	0	0.020	−0.145	−0.203	0.925
17	1	0.965	0.191	0.268	−0.570
18	0	0.752	−1.740	−1.669	−1.863
19	0	0.752	−1.740	−1.669	−1.863
20	1	0.935	0.263	0.366	−0.223
21	1	0.972	0.171	0.239	−0.709
22	0	0.230	−0.547	−0.724	−0.510
23	1	0.998	0.040	0.057	−4.086
24	1	0.935	0.263	0.366	−0.223
25	0	0.013	−0.115	−0.162	1.272
26	1	0.956	0.215	0.301	−0.439
27	0	0.053	−0.236	−0.330	0.336
28	1	0.811	0.483	0.648	0.383
29	1	0.706	0.645	0.834	0.684
30	1	0.656	0.724	0.919	0.809

TABLE 3.3. Residuals for logit model fit to data of Table 3.1. The columns represent (1) student number, (2) observed values (1=pass, 0=fail), (3) fitted probabilities, (4) Pearson residuals, (5) deviance residuals and (6) adjusted deviance residuals. Note the anomalous value of the adjusted deviance residual for student 23, despite the accurate prediction of the observed value.

to evaluate the size of these residuals, due to the non-normality of the sampling distributions of the residuals in this setting.

Bayesian residuals

In contrast to classical residual analysis, Bayesian residual analysis is based on direct examination of the probability distribution of the difference between the observed proportion and the fitted proportion. For binary regression models, Bayesian residuals may be defined as

$$r_{i,B} = y_i/n_i - p_i = \hat{p}_i - F(x_i'\beta), \tag{3.31}$$

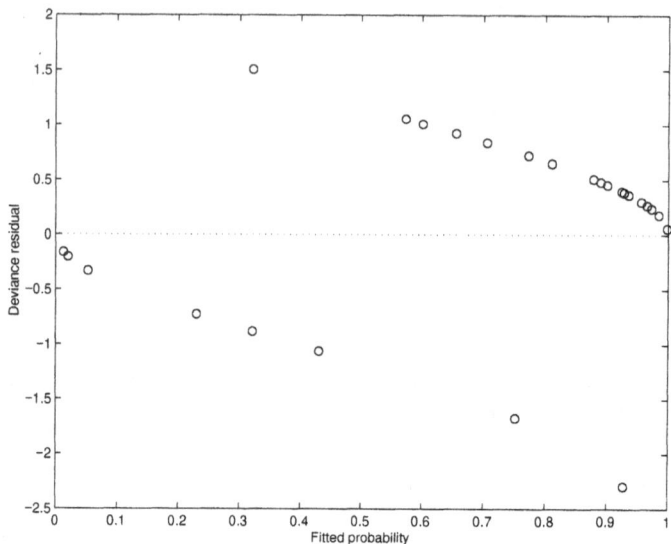

FIGURE 3.7. Deviance residuals versus fitted values for logit regression of pass probability on SAT-M score.

where $\hat{p}_i$ denotes the observed proportion of success for observation i. In this expression, it is important to keep in mind that the only random quantity is β and that the posterior distribution of β determines the posterior distribution of the residual $r_{i,B}$.

Bayesian residual analyses have the important advantage that the distribution of the residuals for small n_i is well defined and can be examined in meaningful ways without resorting to asymptotic arguments. Indeed, for Bernoulli observations, Bayesian residual distributions appear to present the only viable method for identifying outliers.

Although the posterior distributions of the quantities $r_{i,B}$ are not available analytically, histogram estimates of their distributions can be obtained using a MCMC sample from the posterior distribution of the regression parameter β. Denoting these sampled values by $\beta^{(j)}$, $j = 1, \ldots, m$, sampled values from the residual posterior distribution corresponding to observation i are defined by

$$r_{i,B}^{(j)} = y_i/n_i - F(\mathbf{x}_i'\beta^{(j)}), \quad j = 1, \ldots, m. \tag{3.32}$$

The Bayesian residual distributions for the pass/fail model using a logistic link are graphed using line plots against the mean fitted probabilities in Figure 3.8. This plot resembles Figure 3.5, in which plots of the posterior distributions of the fitted probabilities p_i against the SAT-M score were displayed. In Figure 3.8, the line segments indicate the extent of the 5th and 95th percentiles of each residual distribution, whereas the circles indicate the median of each residual's posterior. A residual value of 0 is indicated by the dotted line. Residual distributions located far from the line correspond to possible outliers. Four distributions appear extreme—

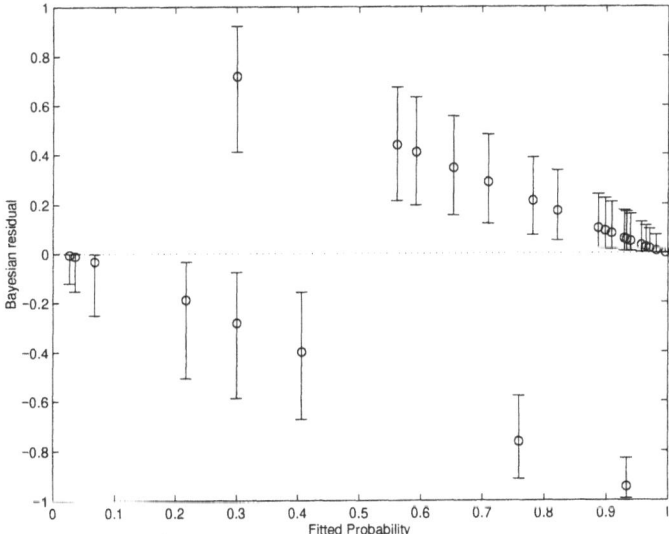

FIGURE 3.8. Bayesian residual distributions graphed against the fitted probabilities for the statistics class dataset. Each line graph shows the location of the 5th, 50th, and 95th percentiles of the posterior distribution of $\hat{p}_i - p_i$.

three with negative residuals (two distributions overlap in Figure 3.8) and one with a positive residual.

Posterior-predictive residuals

Another diagnostic tool can be based on the predictive distributions of observed values and the posterior distribution on the regression parameter. Recall from Chapter 1 that, in binomial models, the predictive distribution $f(y^*)$ is the probability of y^* successes in a future binomial sample of size n^*. This distribution is obtained by averaging the binomial probability of y^* successes, $f(y^*|p)$, over the posterior density on the success probabilities p. Thus, to obtain the posterior-predictive residual for the ith observation, define y_i^* to be the predicted number of successes in a future sample of size n_i, with associated covariate vector x_i, where n_i and x_i are the same sample size and covariate vector corresponding to the ith observation y_i. The predictive distribution of this future observation can be expressed as

$$f(y_i^*|\mathbf{y}) = \int f(y_i \mid \mathbf{x}_i, \beta)g(\beta \mid \mathbf{y})d\beta, \qquad (3.33)$$

where $f(y \mid \mathbf{x}_i, \beta)$ denotes the sampling density of a binomial random variable with sample size n_i and success probability $F(\mathbf{x}_i'\beta)$, and $g(\beta \mid \mathbf{y})$ is the posterior density of the regression vector β based on the observed data $\mathbf{y}$. We refer to this density as the *posterior-predictive* distribution, since one is averaging the predictive binomial density over the posterior distribution of β.

A particular observation may be defined as an outlier if the observed number of successes, y_i, is inconsistent with the corresponding posterior-predictive distribution $f(y_i^*|\mathbf{y})$. In other words, if the observed value y_i falls in the extreme tails of this posterior-predictive distribution, then we may conclude that this observation is not consistent with the assumed model for the data.

In practice, the posterior-predictive distribution $f(y_i^*|\mathbf{y})$ can be computed by simulation. One value of y_i^* can be simulated by the following two-step process:

1. Simulate a sample from the posterior distribution on the regression parameter β. Call the simulated sample $\beta^{(1)}, \dots, \beta^{(m)}$.
2. For each sampled value of β, $\beta^{(j)}$, draw a binomial random variable $y_i^{*(j)}$ with sample size n_i and success probability $p_i^{*(j)} = F(x_i'\beta^{(j)})$.

Based on the simulated binomial random variables, a histogram estimate for the posterior-predictive distribution for each binomial observation can be constructed.

In our pass/fail model, all observations were binary, so $n_i = 1$ for all i. Applying the algorithm described above to obtain samples from the posterior-predictive distribution for the observed values of the covariates is therefore trivial and requires only a bit more computation than required to obtain the posterior distributions on the residuals. Values of the posterior-predictive distributions obtained using this technique are summarized in Table 3.4, along with the corresponding values of $\{y_i\}$. To detect unusual observations, we search for cases in which $\Pr(y_i^* \leq y_i|\mathbf{y})$ or $\Pr(y_i^* \geq y_i|\mathbf{y})$ (the number in the last column) is unusually small. For example, case 4 stands out as unusual, since $y_4 = 0$ and $\Pr(y_4^* = 0|\mathbf{y}) = .068$. Cases 12, 18, and 19 also have relatively small predictive probabilities. These are the same four observations that stood out in the graph of the Bayesian residuals. However, by examining the posterior-predictive distribution for these observations, we can conclude that the residuals are not terribly extreme, and in fact, 0.068 is not an unusually small value for the smallest probability in a set of 30 observations.

An alternative way of viewing the posterior-predictive distribution is to define a posterior-predictive residual as being the difference between the observation and predicted future observation:

$$r_{i,PP} = y_i - y_i^*.$$

Like the Bayesian residuals $r_{i,B}$, the set of residuals $\{r_{i,PP}\}$ have probability distributions which can be graphed as a function of the observation number to learn about goodness of fit. Any residuals which concentrate most of their probability mass away from zero correspond to possible outliers.

Cross-validation residuals

Each of the residuals defined above suffers from an effect called masking. Masking occurs when the apparent size of a residual is decreased due to the fact that the observation itself is used in estimating the value of the regression parameter, which, in turn, is used in the calculation of the residual. In the binary regression setting, an observed proportion that falls well off the regression surface will tend to pull

Student #	y	Posterior-predictive dist. $\Pr(y_i^* = 0)$	$\Pr(y_i^* = 1)$	$\Pr(y_i^* = y_i)$	Cross-val. dist. $\Pr(y_i^* = y_i)$
1	0	0.699	0.301	0.699	0.656
2	0	0.593	0.407	0.593	0.545
3	1	0.408	0.592	0.592	0.557
4	0	0.068	0.932	0.068	0.027
5	1	0.071	0.929	0.929	0.926
6	1	0.092	0.908	0.908	0.904
7	1	0.113	0.887	0.887	0.882
8	1	0.019	0.981	0.981	0.980
9	1	0.218	0.782	0.782	0.770
10	1	0.439	0.561	0.561	0.520
11	1	0.092	0.908	0.908	0.904
12	1	0.699	0.301	0.301	0.193
13	1	0.102	0.898	0.898	0.894
14	1	0.068	0.932	0.932	0.929
15	1	0.102	0.898	0.898	0.894
16	0	0.965	0.035	0.965	0.959
17	1	0.036	0.964	0.964	0.962
18	0	0.241	0.759	0.241	0.191
19	0	0.241	0.759	0.241	0.191
20	1	0.061	0.939	0.939	0.936
21	1	0.030	0.970	0.970	0.969
22	0	0.782	0.218	0.782	0.747
23	1	0.004	0.996	0.996	0.996
24	1	0.061	0.939	0.939	0.936
25	0	0.973	0.027	0.973	0.970
26	1	0.044	0.956	0.956	0.954
27	0	0.933	0.067	0.933	0.922
28	1	0.179	0.821	0.821	0.812
29	1	0.291	0.709	0.709	0.691
30	1	0.347	0.653	0.653	0.628

TABLE 3.4. Posterior-predictive and cross-validation residual distributions for the logit model fit to the data of Table 3.1. The columns represent (1) student number, (2) observed values (1 = pass, 0 = fail), (3) posterior-predictive distribution of the binary outcome, (4) the probability that the future observation is equal to the observed value, and (5) the probability that the cross-validation future observation is equal to the observed value.

the fitted model toward itself. As a result, the distance between the fitted model and the observed proportion will be smaller than the corresponding distance from the "true" regression model.

For classical and Bayesian residuals, masking can be overcome by first refitting the regression surface after omitting the suspicious observation from the data vector and then looking at the difference between the observation and a predicted observation from the fitted model. Let $\mathbf{y}_{(i)}$ denote the vector of observations $\mathbf{y}$ with the ith observation y_i removed, and let $g(\beta|\mathbf{y}_{(i)})$ denote the posterior density of β from the sample of $n - 1$ observations. The *cross-validation* predictive density of y_i^* successes in a sample of size n_i with associated covariate vector x_i is then defined as

$$f(y_i^* | \mathbf{y}_{(i)}) = \int f(y_i^* | \mathbf{x}_i, \beta) \, g(\beta | \mathbf{y}_{(i)}) \, d\beta. \qquad (3.34)$$

To check for outliers, we compare the observation y_i to its corresponding predictive density $f(y_i^* | \mathbf{y}_{(i)})$. This comparison can be made by inspecting the distribution of the cross-validation residuals

$$r_{i,CV} = y_i - y_i^*,$$

where y_i^* is the predicted observation whose distribution is based on the "leave-one-out" posterior distribution of β.

The last column Table 3.4 displays the values of the cross-validation predictive probabilities $\Pr(y_i^* = y_i | \mathbf{y}_{(i)})$, which can be compared with the corresponding posterior-predictive probabilities displayed in the previous column. Note that the four possible outliers (cases 4, 12, 18, and 19) each have smaller predictive probabilities using the cross-validation approach. This suggests that these observations were influential in the fit of the model, and by excluding these observations from the dataset, we would obtain somewhat different estimates of β. Also, the argument that case 4 is an outlier is strengthened using the cross-validation residuals, since the probability of observing this value has now decreased to about 0.03. Still, such a probability is not unexpected in a sample of this size, and so we cannot conclude that this observation is an outlier.

A difficulty with cross-validation residuals is that it may not be computationally feasible to rerun the MCMC algorithm to obtain the posterior distribution on β for each excluded observation. An alternative to rerunning the entire MCMC sampling algorithm to obtain these cross-validation densities is to resample values from the original chain (e.g., Gelfand, 1996). Letting $\mathbf{y}_{(i)}$ denote the data vector $\mathbf{y}$ with the ith observation omitted, we can define resampling weights w_{ij} by

$$w_{ij} = \frac{1}{f(y_i | \mathbf{y}_{(i)}, \beta_j)},$$

where $f(y_i | \mathbf{y}_{(i)}, \beta_j)$ is the predictive density at y_i given β_j. For binomial data in which the binomial observations are independent, $f(y_i | \mathbf{y}_{(i)}, \beta_j)$ is just the contribution to the likelihood function from the single observation y_i, given β_j. Using these resampling weights, histogram estimates of the predictive density of the residuals can be obtained by sampling (with replacement) values of β_j, with weights proportional to w_{ij}. In practice, it is generally preferable to use this resampling scheme to estimate the distribution of the cross-validation residuals instead of resampling the entire chain.

Bayesian latent residuals

The Bayesian residuals $r_{i,B} = \hat{p}_i - p_i$ can be difficult to interpret for a variety of reasons. One problem that arises in interpreting these residuals is the fact that the marginal distributions of the ordered residuals differ. For example, the distribution of the smallest residual is different from that of the median residual. This makes it difficult to assess how extreme each distribution is, while simultaneously accounting for the fact that the most extreme residuals were selected for inspection. In addition, Bayesian residuals do not account for the sample variation of the data, and so determining whether a residual is actually extreme or not can be difficult. Latent residual posterior densities provide relief from this problem.

Bayesian latent residuals are defined using the latent variables described in Section 3.3. In terms of the latent variables $Z_1, ..., Z_n$, we may define latent residuals as

$$r_{i,L} = Z_i - x_i'\beta.$$

Nominally, $r_{1,L}, ..., r_{n,L}$ are a random sample from the distribution F. In the case of a logit model, these residuals are a priori independently distributed from a standard logistic distribution.

A convenient diagnostic tool for examining the joint posterior distribution of the latent residuals is the quantile-quantile plot. A quantile-quantile plot is used to compare ordered sampled values of the latent residuals to the expected order statistics of the link distribution. By comparing ordered samples or expected order statistics to the expected values of the order statistics from the model distribution, one may assess whether the values of large residuals were, in fact, extreme for the given sample size.

In the pass/fail example, the fourth observation has both a large deviance residual and a posterior Bayesian residual density that assigns negligible mass near zero. To examine whether the magnitude of this residual is large after account is made for the number of observations in the study, we sampled 5000 latent residuals for each observation using a Gibbs sampler similar to that described in Section 3.3. A quantile-quantile plot (or logistic scores plot) for the posterior means of the ordered latent residuals is displayed in Figure 3.9. The posterior means were obtained by ordering the sampled latent residuals at each iteration and then averaging the ordered sampled residuals across all iterations. One difficulty in interpreting these posterior means is the loss of identification of the observations in the averaging process—the observation that corresponds to the smallest latent residual may vary over iterations. In particular, note from Figure 3.9 that there are two points at the lower left that fall below the 45° line. However, the smallest ordered residual in the simulations did not always correspond to the same observation. The label on the graph indicates that observation 4 had the smallest latent residual in 61% of the iterations. Also, this observation had the second smallest residual 25% of the time, which indicates that, with probability .86, observation 4 had either the smallest or second-smallest residual. Since both of these residuals fall below the line, this suggests that observation 4 may not conform to model assumptions made for this logistic model.

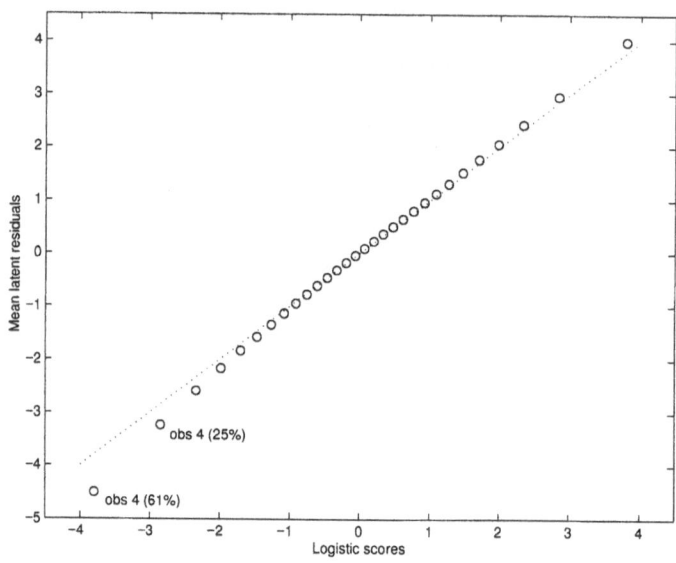

FIGURE 3.9. Logistic scores plot of the posterior means of the sorted latent residuals for statistics class dataset. The labeled points indicate the percentages that particular observations contributed in the computation of the corresponding posterior mean residual.

3.4.2   Goodness of fit and model selection

Classical methods

Standard methods for assessing goodness of fit focus on the deviance statistic. Unlike normal error models in which observational variance is a model parameter, the variance of binomial observations is determined by their mean, and so the deviance statistic often provides an "honest" statistic for assessing goodness of fit and model comparison.

Recall that the deviance for measuring the agreement of a binary regression model with observed data is given by

$$D = 2 \sum_{i=1}^{n} \left\{ y_i \log \left(\frac{y_i}{\hat{y}_i} \right) + (n_i - y_i) \log \left(\frac{n_i - y_i}{n_i - \hat{y}_i} \right) \right\}, \tag{3.35}$$

where $\hat{y}_i = n_i \hat{p}_i$ is the fitted value from the model. If the model is "true," the deviance follows an asymptotic χ^2 distribution with $n - q$ degrees of freedom, where n is the number of binomial observations and q is the dimension of the regression parameter. Under the assumption of independent binomial observations, this result follows from the relationship of the deviance statistic to the likelihood ratio statistic and holds as $n_i \to \infty$, with n fixed. Unfortunately, for Bernoulli observations and other cases where the binomial sample sizes n_i are small, the asymptotic χ^2 distribution of the deviance statistic may not pertain. Indeed, for linear logistic models with Bernoulli observations, the deviance function can be expressed solely as a function of the MLE of the regression parameter, which demonstrates the futility of using this statistic to measure goodness of fit for such data.

For purposes of model comparison, deviance statistics from several competing models are often displayed in the form of an analysis of deviance (AOD) table. AOD tables are used to display models along with their associated degrees of freedom $(M - q)$ and deviance. Nested models within an AOD table can be compared using their deviance and degrees of freedom. For example, consider the case of two models M_1 and M_2, where M_1 is nested within M_2 in the sense that all of the parameters in model M_1 are also included in model M_2. Suppose that the simpler model, M_1, has deviance D_1 and df_1 degrees of freedom, and model M_2 has deviance D_2 on df_2 degrees of freedom. The difference of deviance values, $D_1 - D_2$ is nominally distributed as a χ^2 random variable with $df_1 - df_2$ degrees of freedom. The simpler model M_1 is then rejected if the difference in deviance is large relative to the expected value of a $\chi^2_{df_1 - df_2}$ distribution. A balance between parsimony and fit is obtained by selecting models in which exclusion of any explanatory variable results in a comparatively large increase in deviance.

To illustrate the use of AOD tables in model comparison, we again consider the data of Table 3.1. In the logistic model presented above for this data, the grades of the students in the prerequisite probability class were ignored. However, the grade in this class might reasonably be thought to predict success in the statistics class, and so this explanatory variable might also be considered for inclusion in any model for the pass probabilities.

The prerequisite grade can be incorporated into the logistic regression model in several ways. One possibility for its inclusion is to create a factor variable in which each grade level is assumed to have an effect that is estimated independently for each grade. This factor model may be specified as

$$\log[p_i/(1 - p_i)] = \begin{cases} \beta_1 \times \text{SAT-M}_i + \beta_2 & \text{if grade is A} \\ \beta_1 \times \text{SAT-M}_i + \beta_3 & \text{if grade is B} \\ \beta_1 \times \text{SAT-M}_i + \beta_4 & \text{if grade is C} \\ \beta_1 \times \text{SAT-M}_i + \beta_5 & \text{if grade is D} \\ \beta_1 \times \text{SAT-M}_i + \beta_6 & \text{if grade is F.} \end{cases} \quad (3.36)$$

In this model, β_2 represents the effect of receiving an A on the logistic probability of passing, β_3 the effect of receiving a B, and so on. Note that a constant term is not included in this model. If we included a constant term along with a dummy variable for each grade received in the prerequisite course, the model would be overparameterized and we would not obtain unique estimates of the regression coefficients.[1]

---

[1] Note also that the maximum likelihood estimate of the regression coefficient associated with a grade of F in the prerequisite probability course is infinite in this example, due to the fact that the only student who received an F in the probability course passed the statistics class. When examining computer output summarizing the maximum likelihood

A more parsimonious model that might be used to incorporate the effects of the previous course's grade could be formulated by assuming an interval effect on the success probabilities from these grades. An obvious way to model this type of effect would be to use the standard GPA coding. In other words, A's would be assigned the value 4.0, B's 3.0, ..., and F's the value 0. Denoting the coded values of the grades by $GRADE_i$, we obtain a model of the form

$$\log\left(\frac{p_i}{1 - p_i}\right) = \beta_0 + \beta_1 \times \text{SAT-M}_i + \beta_2 \times \text{GRADE}_i. \qquad (3.37)$$

An AOD table comparing these three models and the constant model with no covariates is provided in Table 3.5. To see if the SAT variable should be included in the model, we compare the constant and SAT models in the table. The difference in deviance between the two models is $36.65 - 22.27 = 13.38$ with $29 - 28 = 1$ degrees of freedom. The value of 13.38 is significant for a χ^2 variable on 1 df, so, based on the AOD table, we conclude that the SAT score cannot be removed from the model. Next, note that model (3.21) is nested in both models (3.36) and (3.37), and so the likelihood ratio test for comparing model these models to (3.21) is just the difference in the deviance between models. In comparing model (3.21) to model (3.36), the test statistic has approximately a χ^2 distribution on $28 - 24 = 4$ degrees of freedom. The statistic for comparing model (3.21) to model (3.37) is nominally χ^2 on $28 - 27 = 1$ degrees of freedom. In both cases, the value of the test statistic is not significant for the associated chi-squared distribution and so the simpler model which includes only the SAT variable is preferred. In other words, the additional information contained in the previous course's grade appears not to be significant.

Bayesian methods

The Bayesian approach to comparing models is based on the notions of marginal likelihood and Bayes factors. We introduce these ideas in the context of binary regression models, although the methods are applicable to a wide variety of model comparisons. It should be stated at the outset, however, that formal Bayesian model

Model	Covariates	Deviance	Degrees of freedom
—	None	36.65	29
(3.21)	SAT-M	22.27	28
(3.37)	SAT-M, grade as interval covariate	21.95	27
(3.36)	SAT-M, grade as factor variable	18.92	24

TABLE 3.5. Analysis of deviance table for pass/fail data.

---

fitting procedure, either unusually large values of a regression coefficient or large values of a standard error may be used to diagnose the occurrence of such values.

selection procedures require the specification of proper prior[2] densities on all parameters of interest.

Suppose then that we observe data $\{y_i, n_i\}$, where, as before, y_i represents the number of successes in a sample of size n_i with underlying probability of success p_i. The probabilities are linked to a linear regression model by the equation $p_i = F(x_i'\beta)$, where the cdf F is known and the regression parameter β is unknown. A *Bayesian model M* for this problem is comprised of the specification of a sampling density for **y** and a proper prior density for β. For example, model M_1 might specify that the y_i are independent binomial observations, F is a logistic distribution, and β has a normal distribution with known mean and covariance matrix. A second model, M_2, might posit independent binomial distributions for y_i, but might, instead, assume a complementary log-log link function, with β distributed a priori as a multivariate t distribution with known degrees of freedom, mean, and covariance matrix.

In general, if we let $f(\mathbf{y}|\beta)$ denote the sampling density of **y** and let the prior density on the unknown model parameters be denoted by $g(\beta)$, then the joint density of the data and parameter is equal to the product of $f(\mathbf{y}|\beta)$ and $g(\beta)$. The Bayesian measure of the goodness of fit of model M is given by the *marginal likelihood* of the data which is obtained by integrating the parameter β from the joint density:

$$f(\mathbf{y}|M) = \int f(\mathbf{y}|\beta)g(\beta)d\beta.$$

We write the marginal likelihood of the data as $f(\mathbf{y}|M)$ to emphasize its dependence on the particular choice of model. Note that this quantity is equivalent to the prior predictive density obtained by integrating the binomial density $f(\mathbf{y}|\beta)$ over the prior density $g(\beta)$.

Within a testing context, we are often faced with the problem of choosing between two models. If the two models are denoted by M_1 and M_2, then the measure of evidence in support of model M_1 over M_2 is given by the *Bayes factor*, defined as the ratio of the marginal likelihoods of the two models:

$$\mathrm{BF}_{12} = \frac{f(\mathbf{y}|M_1)}{f(\mathbf{y}|M_2)}.$$

We cite Bayes factors on the log 10 scale. Thus, if $\log \mathrm{BF}_{12} = -2$, then model 2 is 100 times more likely than model 1, and if $\log \mathrm{BF}_{12} = 0$, the two models are equally supported by the data.

To illustrate the use of Bayes factors, let us first consider the question of whether the SAT-M score should be included in the logit model predicting success in the statistics class. To compute the Bayes factor for this test, we need to construct the prior distributions for the regression vector β under two models: one model

---

[2] A proper prior density is a prior density that integrates to one. A uniform prior on real-valued parameter, like a regression parameter, is not proper because its integral over the real line is infinite.

which assumes that SAT-M does affect the pass probability and a second model which assumes that SAT-M is not a useful predictor. The prior for each model can be constructed using the conditional means methodology introduced in Section 3.2.3. If we believe that SAT-M does not predict success in the class, then two SAT-M scores and two probability guesses can be chosen that reflect this opinion. For example, these beliefs can be quantified by supposing that the probability of passing is .7 for a student with a 500 SAT-M score and the probability is also .7 for a student with a 600 score. (The observed passing rate in the class was 70%—we are intentionally weighting the priors to provide the most favorable evidence in favor of the "null" hypothesis of no effect.) To complete this prior specification using the conditional means approach, we have to state the number of prior observations that this information is worth. If SAT-M should be included in the model, then these prior statements should have little influence on the corresponding posterior distribution—thus we assign only a single observation to this belief. On the other hand, if SAT-M is not a useful predictor and should be removed from the model, then we want to assign a prior which will be very influential and force the posterior distribution to remove this term from the model. To make this prior information influential, we assign a very large number of observations, say 1000, to the belief that SAT-M does not affect the pass probability.

To summarize, our two prior distribution models are derived from the following elicitations. As in Section 3.2.3, g denotes a prior guess at the pass probability and K represents the number of observations that the guess is worth. In all cases, this information is matched to a beta prior density with parameters Kg and $K(1-g)$. In the construction of the conditional means prior, the densities of the two pass probabilities for different covariate values are assumed to be independent.

- Model M_1 (SAT-M should be included in model): When SAT-M $= 500$, $g = .7$; when SAT-M $= 600$, $g = .7$; $K = 1$.
- Model M_2 (SAT-M should be removed from model): When SAT-M $= 500$, $g = .7$; when SAT-M $= 600$, $g = .7$; $K = 1000$.

Having completed the model specifications, we next compute the marginal likelihood for each model. The ratio of these marginal likelihoods, the Bayes factor, is a measure of the goodness of fit of the model which includes the SAT-M term. In this case, the marginal likelihoods are $\log f(y|M_1) = -21.07$ and $\log f(y|M_2) = -24.73$. Thus, the Bayes factor in support of SAT-M being included in the model is

$$BF_{12} = \frac{f(y|M_1)}{f(y|M_2)} = 38.6.$$

On a log 10 scale, the Bayes factor is 1.6. This indicates that there is significant evidence for the inclusion of SAT-M in the model.

One concern in the computation of the Bayes factor is the choice of the parameter K in the specification of the two prior models. Specifically, what is the effect of this parameter on the value of the Bayes factor? Table 3.6 answers this question by tabulating values of the log marginal likelihood for a range of values of $\log_{10} K$.

$\log_{10} K$	$\log f(\mathbf{y}|K)$
-2	-27.91
-1	-23.59
0	-21.07
1	-22.60
2	-24.38
3	-24.73
4	-24.76
5	-24.77

TABLE 3.6. Values of the logarithm of the marginal likelihood for values of the prior information parameter K. In all cases, a conditional means prior is used which states that $g = .7$ when SAT-M = 500, and $g = .7$; when SAT-M = 600. Independent beta priors are used to match this prior information.

Note that the value of $\log f(\mathbf{y}|M)$ is very stable for values of $\log_{10} K$ equal to 3 or greater. The marginal likelihood for the value $K = 1000$ that we used in the example is approximately equal to the marginal likelihood obtained by letting the number of prior observations K approach infinity. Second, note that the value of the logarithm of the marginal likelihood decreases rapidly as the parameter K decreases toward zero. In fact, the marginal likelihood is not defined if K is chosen to be zero, since this would correspond to an improper prior. The value $K = 1$ represents a modest amount of prior information which will have little influence on the posterior distribution but result in a stable value of the marginal likelihood.

Given that SAT-M is to be included in the regression model, we might next be interested in testing whether a student's grade in the prerequisite class should also be added. As before, we assume that grade may be coded as an interval variable with A=4, B=3, and so on. The null model in this case, say model M_1, only includes the term SAT-M; this model states that the coded grade is not a useful predictor of the pass probability. This implies that students with identical SAT-M scores and different prerequisite grades should have identical probabilities of passing the course. As above, we quantify this knowledge by assigning prior guesses at the probability of success for two students with the same SAT-M but different prerequisite course grades. Thus, we assume that the probability of passing for two students with covariates (550, B) and (550, D) is 0.67, the probability predicted by the model including just SAT-M. Once again, we assume a high degree of certainty in this belief by taking $K = 1000$. By assuming this large number of prior observations, the posterior distribution effectively removes the grade covariate from the model.

In the alternative model, say M_2, we assume that the prerequisite course grade improves the prediction of the pass probability, although we are uncertain as to the precise magnitude of this effect. To model this state of knowledge, we again guess that the pass probability for two students having covariate values (550, B) and (550, D) is 0.67, but now assume that our prior certainty regarding this probability is low. We reflect this uncertainty by taking $K = 1$. This small value of K permits the inclusion of the prerequisite course grade in the model.

Since there are two covariates in the proposed model, a proper prior will consist of a conditional means prior with three independent components. Since only two prior components are stated in models M_1 and M_2 above, we complete the prior specification by assuming for each model that the probability that a student passes with covariate value $(500, C)$ is .1 and this information is worth $K = 1$ observation. This third prior component is chosen only to complete the prior specification for each model and has little impact on the Bayes factor computed below.

These two models can be summarized given as follows:

- Model M_1 (GRADE should be removed from model): When SAT-M $= 550$ and GRADE $= B$, $g = .67$, $K = 1000$; when SAT-M $= 550$ and GRADE $= D$, $g = .67$, $K = 1000$; when SAT-M $= 500$ and GRADE $= C$, $g = .1$, $K = 1$.
- Model M_2 (GRADE should be included in model): When SAT-M $= 550$ and GRADE $= B$, $g = .67$, $K = 1$; when SAT-M $= 550$ and GRADE $= D$, $g = .67$, $K = 1$; when SAT-M $= 500$ and GRADE $= C$, $g = .1$, $K = 1$.

A Bayesian test for the inclusion of GRADE (assuming SAT-M is in the model) is based on comparing models M_1 and M_2 by means of a Bayes factor. In this case, $\log f(y|M_2) = -22.72$ and $\log f(y|M_1) = -20.16$. Thus,

$$\mathrm{BF}_{12} = \frac{f(y|M_1)}{f(y|M_2)} = .0776.$$

On a log 10 scale, the Bayes factor in support of the inclusion of GRADE is -1.11. Since this is negative, this test is supportive of the simpler model which only includes the SAT-M score.

Comparing the results of these Bayesian tests with the classical deviance tests, note that similar conclusions were reached by the two tests about the significance of the SAT-M and prerequisite grade covariates. One advantage of the Bayesian approach is that it can provide support for simpler models containing fewer covariates. In contrast, the deviance always decreases as covariates are added to models, and so the deviance ratio test will always tend to favor larger models.

3.5   An example

We now turn attention to a more realistic example. The subject of our attention is a study conducted by Dr. Robert Terry and colleagues at Duke University involving conduct disorders of "high-risk" elementary and secondary school children in the Durham public school system. This ongoing longitudinal study was designed to assess, among other things, the effects of social rejection and aggression in early schooling on later incidence of conduct disorder. In the subset of data we consider, binary variables indicating aggressive behavioral patterns and social rejection were collected for 172 third-graders. The same children were followed through high school and monitored for conduct disorders in grades 6, 8, 10 and 12. The gender of each student is also available. The subset of data examined here represents a small portion of the data collected during the study, and for simplicity, we have excluded all cases with missing explanatory or response variables.

Our response variable **y** is the number of grades at which each student exhibited "conduct disorder." Because conduct disorder was recorded in grades 6, 8, 10 and 12, the possible values of our response are 0, 1, 2, 3, and 4. Dichotomous explanatory variables for the ith individual include sex (S_i; 0–male, 1–female), aggressive behavior in the third grade (A_i; 1 if observed, 0 otherwise), and social rejection in the third grade (R_i; 1 if rejected, 0 otherwise). We initially ignore possible serial correlation in the occurrence of conduct disorder within individuals and assume as a baseline model a binomial distribution for each component of **y**; that is, we assume $y_i \sim \text{Bin}(4, p_i)$.

An exploratory look at the data reveals that the mean number of conduct disorder diagnoses for males was 1.64, and for females 1.51. Among those children exhibiting aggressive behavior in the third grade, the mean response was 1.93; for those not exhibiting such behavior, the mean number of conduct disorder diagnoses was 1.50. For socially rejected third-graders, the mean was 1.49; for nonsocially rejected third-graders it was 1.61.

Based on this preliminary examination of the data, a logistic regression model of the form

$$\log\left(\frac{p_i}{1 - p_i}\right) = \beta_0 + \beta_S S_i + \beta_A A_i + \beta_R R_i \qquad (3.38)$$

was posited for the data. The maximum likelihood estimates and asymptotic standard deviations for the components of β appear in Table 3.7. The deviance for this model was 340 on 168 degrees of freedom. Although the asymptotic χ^2_{168} distribution is unlikely to pertain in this setting due to the fact that each binomial sample size is 4, the large value of this statistic probably indicates some lack of model fit.

A plot of deviance residuals versus fitted values appears in Figure 3.10. Many of the deviance residuals in the plot appear suspiciously large, but as in the case of the deviance statistic itself, the magnitudes of these residuals are difficult to gauge due to the relatively small binomial sample sizes.

To more accurately assess the fit of this model, we next performed a Bayesian analysis. For simplicity, we assumed a vague prior for the components of β and obtained posterior samples for the regression coefficients and residuals using the Metropolis-Hastings algorithm described in Section 3.2.3.

Posterior means and standard deviations obtained from the Metropolis-Hastings algorithm appear in Table 3.8. These values are consistent with the maximum likelihood estimates summarized in Table 3.7 and indicate that the asymptotic normal approximation provides a reasonable summary of the posterior distribution.

The Bayesian methodology developed in Section 3.4 is also useful in performing case analyses. As a first step in this direction, we examined the posterior distribution of the Bayesian residuals. Line plots of these posterior distributions are graphed against observation number in Figure 3.11. In contrast to the pass/fail grading example, the regression coefficients are more precisely known in this case, and

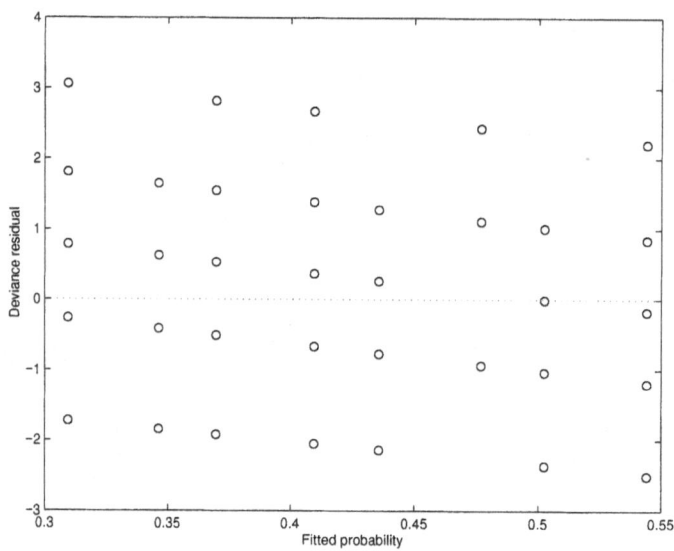

FIGURE 3.10. Deviance residuals versus fitted values for logistic model for conduct disorder. Due to the limited number of explanatory variable combinations, many of the residual values are not distinguishable from one another in the plot.

Parameter	MLE	Standard error
β_0	−0.37	0.12
β_S	−0.17	0.16
β_A	0.54	0.22
β_R	−0.27	0.19

TABLE 3.7. Maximum likelihood estimates for baseline logistic model for conduct disorder.

Parameter	Posterior Mean	Standard error
β_0	−0.36	0.12
β_S	−0.17	0.16
β_A	0.55	0.21
β_R	−0.26	0.19

TABLE 3.8. Bayesian posterior means of regression coefficients in logistic regression model for conduct disorder using a uniform prior for β.

so the posterior distributions of the Bayes residuals are more concentrated. As a result, many of these distributions are concentrated away from zero.

A difficulty with Figure 3.11 is that it ignores the sampling variation in the binomial observations; that is, as more data are collected, the values of the regression parameters becomes increasingly precise, and the distribution of $\hat{p}_i - p_i$ concentrates on the difference between the observed and predicted proportion. Thus, to more fairly assess which observations are outliers, we might also wish to examine the difference in the observed values of y and the posterior-predictive sampling

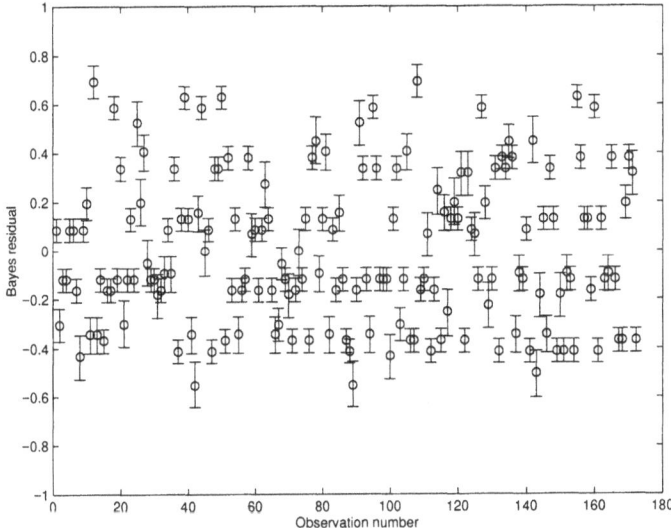

FIGURE 3.11. Bayesian residual distributions graphed against the fitted probabilities for conduct disorder dataset. Each line graph shows the location of the 5th, 50th, and 95th percentiles of the posterior distribution of $\hat{p}_i - p_i$.

distribution of a binomial random variable with a success probability distributed according to $F(\mathbf{x}_i'\beta)$ and sample size n_i. This distribution was defined in (3.33), and the associated random variable is by denoted y_i^*.

Figure 3.12 depicts the interquartile ranges for the distribution of $y_i - y_i^*$. (When viewing this graph, remember that the predictive distribution is concentrated on the five discrete values 0, 1, 2, 3, and 4.) As the figure illustrates, nearly one-half of these interquartile ranges do not include the value 0, and many of the interquartile ranges are concentrated above 2 or less than -2. This fact provides further evidence for model lack-of-fit.

An additional diagnostic based on the posterior-predictive distributions can also be defined by examining the sample variance of the vector $\mathbf{y}$, which is equal to $\sum(y_i - \bar{y})^2/(n-1)$, and the distribution of the variance of sampled values of $\mathbf{y}^*$. These distributions account for both the uncertainty in the predicted success probabilities from the logistic regression model and the binomial variability associated with the observations themselves. In this case, the standard deviation of $y_1, ..., y_n$ is 1.24. To compare this value to the standard deviation of the predicted-posterior distribution on $\mathbf{y}^*$, we simulated samples of $\mathbf{y}^*$ from the predictive distribution, and for each sampled value, we computed the standard deviation of $\{y_i^*\}$. The histogram of standard deviations from the posterior-predictive values is displayed in Figure 3.13. The value of the standard deviation from the observed data is indicated by the vertical line. Note that the observed value is larger than all of the values sampled from the predictive distribution, indicating that the model did not accurately account for the observed variability in the data.

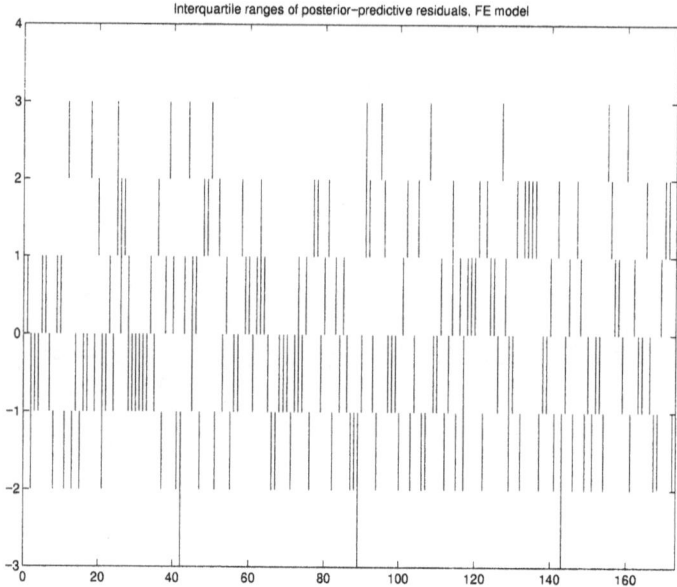

FIGURE 3.12. Interquartile ranges of posterior-predictive residual distributions. Each vertical line depicts the interquartile range for the ith observation, with i plotted on the horizontal axis.

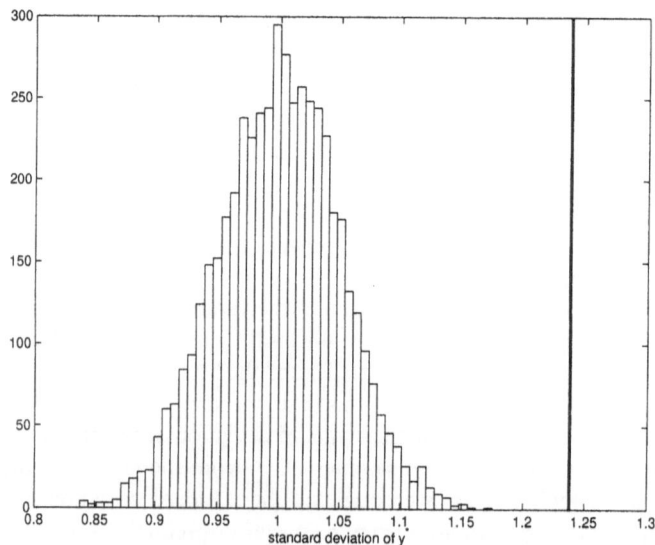

FIGURE 3.13. Histogram of standard deviations of $\{y_i^*\}$ from repeated sampling from the posterior-predictive distribution for logistic model. Observed value of the standard deviation of y is indicated by a vertical line.

To understand the unexpectedly large value of the deviance statistic and the failure of the logistic model to accurately predict the variability in this data, we might consider expanding our model so that it contained parameters that explicitly accounted for sources of additional variation. For these data, one might expect that the observed periods at which conduct disorder was recorded for the same individual might be correlated, even after account is made for values of the explanatory variables. One mechanism for modeling this correlation is to introduce a "random effect" for each individual. This random effect accounts for an individual's propensity toward conduct disorder over that specified by the "fixed effects" of gender, social rejection, and aggressiveness.

One way individual effects on success probabilities can be modeled is through the introduction of normally distributed random effects. For identifiability, the mean random effect is assumed to be 0 and the variance of the random effects is estimated from the data. These considerations lead to a model of the form

$$\log\left(\frac{p_i}{1-p_i}\right) = \beta_0 + \beta_S S_i + \beta_A A_i + \beta_R R_i + \zeta_i, \qquad \zeta_i \sim N(0, \sigma^2). \quad (3.39)$$

Unlike the prior for β, the prior density specified for the random effects variance σ^2 is important. If a prior proportional to $(\sigma^2)^{-p}$, $p \geq 0$, is employed, the posterior distribution will be unbounded (and perhaps non-integrable) when all random effects and σ^2 approach zero. Alternatively, if too much prior weight is assigned to large values of σ^2, the posterior distribution may concentrate on large values of this parameter, allowing the random effects to provide an exact fit of the model to the data. Therefore, some care must be taken in specifying the prior distribution for σ^2. A flexible family of prior distributions for this variance term is the inverse-gamma family of density functions. These densities my be written in the form

$$h(\sigma^2) = \frac{\lambda^\alpha}{\Gamma(\alpha)}(\sigma^2)^{-(\alpha+1)} \exp\left(-\frac{\lambda}{\sigma^2}\right), \qquad \sigma^2 > 0. \quad (3.40)$$

In this application, we use this density with parameters $\alpha = 5$ and $\lambda = 1.5$. This particular density was chosen so that the prior concentrates nearly all of its mass between 0.1 and 1.5.

Sampling from the posterior distribution of parameters in random effects models like (3.39) is more complicated than it is from fixed effect models. This fact is attributable to the increased dimension of the parameter vector and the corresponding decrease in probability of jumps to candidate points. This occurs because "good" jumps in all parameter directions are unlikely to be obtained in random

Parameter	Posterior mean	Standard error
β_0	−0.44	0.18
β_S	−0.17	0.23
β_A	0.63	0.31
β_R	−0.34	0.28

TABLE 3.9. Posterior means of regression coefficients in random effects model for conduct disorder.

draws from the proposal density. To overcome this difficulty, the basic Metropolis-Hastings algorithm can be applied toward sampling the fixed effect parameters, holding random effects and σ^2 fixed and then sampling individual random effects separately. In calculating the posterior density at the candidate and current values of a new random effect value, only the contribution from the given observation need be considered; that is, only the likelihood for the given observation and the prior for that random effect has to be computed. Sampling from the conditional distribution of the variance parameter σ^2 is trivial if a conjugate inverse gamma prior density is selected. In that case, the conditional posterior density of σ^2 also has an inverse-gamma distribution.

Posterior means and standard deviations for the parameters of model (3.39) are provided in Table 3.9. A histogram estimate of the posterior density of the random effects variance σ^2 is displayed in Figure 3.14. This posterior density is concentrated away from zero, supporting the notion that the additional variation in the data may be suitably modeled through the random effects assumption. Further support for the random effects model is provided in the posterior-predictive residual plot, illustrated in Figure 3.15 which should be contrasted to Figure 3.12. The posterior-predictive residual distributions in this plot tend to place significant mass on the point 0, as demonstrated by the fact that more than one-half of the interquartile ranges shown in this plot include 0.

Finally, we assessed the suitability of the random effects model by means of the posterior-predictive distribution. As in the case of the fixed effects model, values of future samples y^* were simulated from the predictive distribution and

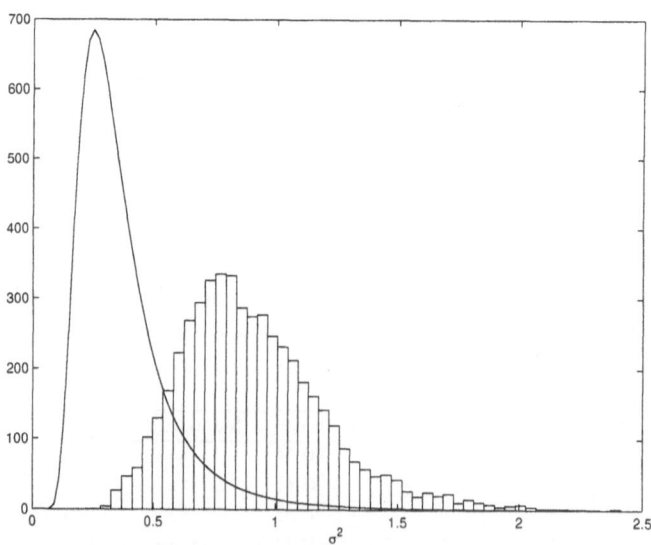

FIGURE 3.14. Histogram summary of sampled values of σ^2, the random effects variance parameter. The line depicts the (rescaled) prior density.

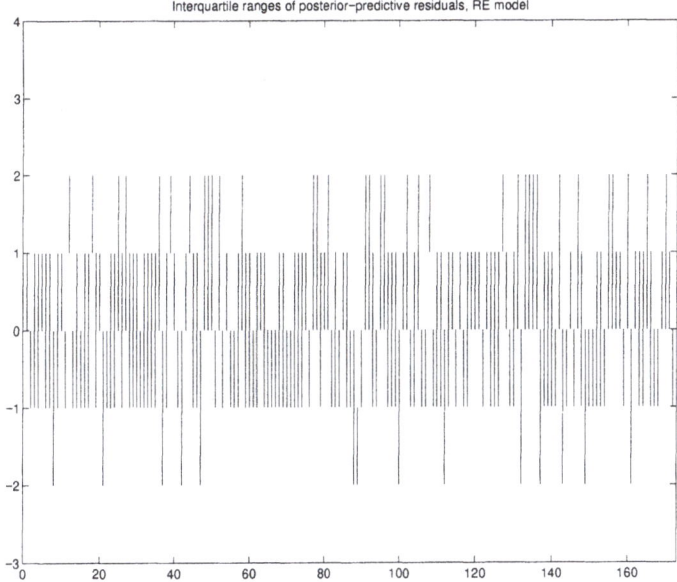

FIGURE 3.15. Interquartile ranges of posterior-predictive residual distributions for the random effects model. Each vertical line depicts the interquartile range for the ith observation, with i plotted on the horizontal axis.

a value of the standard deviation was computed for each simulated sample. The histogram of standard deviations is displayed in Figure 3.16 with the observed standard deviation value of y shown by a vertical line. The observed value is in the center portion of the posterior-predictive distribution, indicating that this value is consistent with the fitted random effects model.

It is interesting to note that the inclusion of the random effects in the Bayesian model for the disorder data leads to conclusions similar to those obtained though a classical analysis which is adjusted for "overdispersion." Typically, the classical correction for overdispersion is obtained by dividing the deviance by the degrees of freedom, in this case $340/168 = 2.02$. Perhaps not surprisingly, the Bayesian random effects model results in approximately the same inflation factor on the posterior variances of the regression parameters.

3.6 A note on retrospective sampling and logistic regression

Throughout this chapter, we have explored relationships among binary random variables, success probabilities, and explanatory variables. In this discussion, we assumed that our data were collected so that the relationships between variables in the dataset were representative of the same relationships in the population at

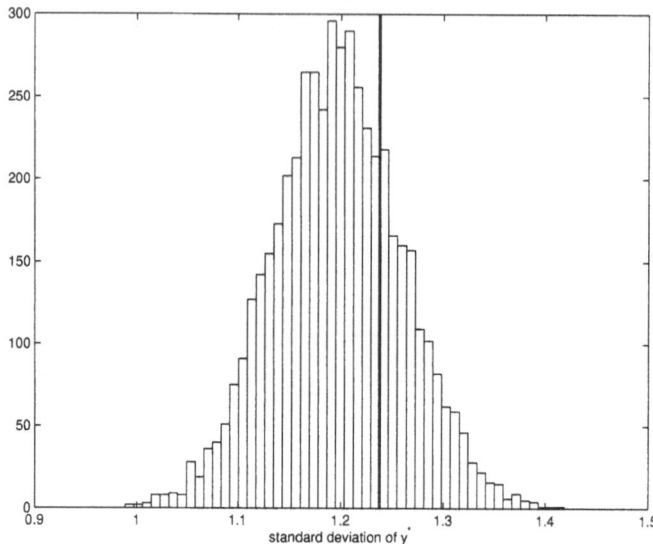

FIGURE 3.16. Histogram of standard deviations of $\{y_i^*\}$ from repeated sampling from the posterior-predictive distribution for logistic random effects model. Observed value of the standard deviation of y is indicated by a vertical line.

large. Indeed, we often tacitly assumed that the items in our data represented a *random* sample from some larger population of interest. Unfortunately, in most observational studies, these assumptions are not correct.

Two observational study designs are commonly used to collect binary data. In the first, a prospective study, items or individuals are selected at random from a population of interest and then categorized as either a success or failure. At the same time, characteristics of the item thought to influence its success or failure are also recorded for subsequent use in the types of analyses described above. In many ways, prospective study designs are optimal because interpretation of statistical analyses based on prospective data is relatively straightforward.

Disadvantages of prospective study designs are that they can be inefficient when successes are rare, and can be very expensive if it is difficult to determine whether an item is a success or failure. For example, in studying variables associated with death due to lung cancer, it would be very costly to randomly select individuals from a given population and to then follow these individuals to determine their cause of death. Furthermore, because lung cancer is a comparatively rare disease, extremely large sample sizes would be required to observe even a modest number of "successes."

Because of the cost and inefficiency associated with prospective studies, retrospective study designs are often used instead. In a retrospective study design, investigators identify successes from some abnormally successful segment of the population and then attempt to choose failures from a similar segment of the population. The items identified as successes are referred to as "cases," whereas failures included in the study for purposes of comparison are called "controls." For obvious reasons, this type of study is also called a case-control study.

Returning to the example of death due to lung cancer, a retrospective study design might involve searching several years of patient records at a hospital to identify patients who died from lung cancer, recording possible covariates from these records and then somehow selecting other patient records to serve as controls. In this way, a large number of patients who died from lung cancer can be studied in a short period of time and at small cost. Of course, collecting data in this way complicates statistical analyses, which must account for the nonrandom sampling mechanism used to obtain the data.

To quantify the effect of the sampling mechanism, let us introduce a variable S_i that indicates whether or not the ith item in a population is sampled, with the convention that $S_i = 1$ if the ith item is sampled, and 0 otherwise. If data are collected using a retrospective sampling design, the probability that $S_i = 1$ depends on the value of y_i, the Bernoulli variable that indicates whether the ith item was a success.

In analyzing data from a retrospective study, we cannot directly model the unconditional probability that $y_i = 1$, given only the vector of covariates x_i' associated with the ith item. Instead, we must model the probability that the ith item is a success, given the vector of covariates x_i and the fact that the item was sampled in the retrospective design. Fortunately, the connection between these two probabilities has a known form when a logistic regression model describes the relation between y_i and x_i.

Assuming a logistic model for y_i and x_i, the probability that we wish to predict is

$$Pr(y_i = 1 \mid x_i) = \frac{\exp(x_i'\beta)}{1 + \exp(x_i'\beta)}. \tag{3.41}$$

However, with data collected using a retrospective study design, the probability that we are actually able to predict is

$$Pr(y_i = 1 \mid x_i, S_i = 1). \tag{3.42}$$

Using Bayes theorem, we can relate the latter probability (3.42) to the desired function in (3.41) as follows:

$$Pr(y_i = 1 \mid x_i, S_i = 1) = \frac{Pr(S_i = 1 \mid y_i = 1, x_i) \, Pr(y_i = 1 \mid x_i)}{Pr(S_i = 1 \mid x_i)}. \tag{3.43}$$

However, the denominator in (3.43) can be reexpressed as

$$Pb(S_i = 1 \mid x_i) = Pr(S_i = 1 \mid y_i = 1, x_i) \, Pr(y_i = 1 \mid x_i)$$
$$+ Pr(S_i = 1 \mid y_i = 0, x_i) \, Pr(y_i = 0 \mid x_i). \tag{3.44}$$

We now make the critical assumption that the probability that an item was sampled is independent of the value of x_i. Assuming that this condition holds, let the probability that a successful item ($y_i = 1$) was sampled be denoted by q, and let the probability that an unsuccessful item ($y_i = 0$) was sampled be denoted by

r. Then, from (3.41) and (3.44), we find that

$$Pr(S_i = 1 \mid \mathbf{x}_i) = q \ \frac{\exp(\mathbf{x}_i'\beta)}{1 + \exp(\mathbf{x}_i'\beta)} + r \ \frac{1}{1 + \exp(\mathbf{x}_i'\beta)}. \qquad (3.45)$$

Noting that the numerator in (3.43) equals the first term on the right-hand side of (3.45), it follows that

$$Pr(y_i = 1 \mid \mathbf{x}_i, S_i = 1) = \frac{q \ \exp(\mathbf{x}_i'\beta)}{r + q \ \exp(\mathbf{x}_i'\beta)}$$

$$= \frac{\exp(\log(q/r) + \mathbf{x}_i'\beta)}{1 + \exp(\log(q/r) + \mathbf{x}_i'\beta)}. \qquad (3.46)$$

Comparing (3.46) to (3.41), we see that the regression coefficients in the two models are identical, except for the intercepts. If the intercept term in (3.41) is β_0, then the intercept in (3.46) is $\beta_0 + \log(q/r)$. Thus, the fact that data were collected retrospectively can be ignored in logistic regression analyses, provided that the linear predictor contains an intercept term and that the sampling mechanism was independent of the explanatory variables. With the exception of the intercept, the coefficients in the regression model have the same meaning as they do in a prospective study. The invariance of the regression coefficients to retrospective sample designs is a property unique to the logistic link; similar results are not applicable to the probit and complementary log-log links.

3.7   Further reading

Binary regression models may be considered as a special case of the class of generalized linear models, which are developed in more generality by Dobson (1990) and McCullagh and Nelder (1989). Classical binary regression models and related diagnostics are detailed in Collett (1991). Dellaportas and Smith (1993) and Albert and Chib (1993) describe MCMC algorithms for fitting Bayesian binary regression models. Chaloner and Brant (1988) and Chaloner (1991) illustrate the use of real-valued Bayesian residuals in normal and exponential regression. The use of latent residuals in model criticism is described in Albert and Chib (1995); Gelman, Carlin, Stern, and Rubin (1995) illustrate the use of posterior-predictive distributions in model checking. Bayes factors are reviewed by Kass and Raftery (1995).

3.8   Appendix: iteratively reweighted least squares

The difficulty in obtaining maximum likelihood estimates for a binary regression model stems from the complexity of (3.16), which makes an analytic expression for the maximum likelihood estimates of β impossible to obtain. However, the iteratively reweighted least squares algorithm makes point estimation of maximum likelihood estimates a trivial task and underlies the algorithm used in most commercial software packages. The same algorithm provides the basis for the MATLAB routines provided on our ftp site, and a rudimentary understanding of the algorithm allows one to modify these routines to accommodate nonstandard link functions and other unusual model features.

For binary regression models, iteratively reweighted least squares proceeds by regressing an adjusted dependent variate z_i on the regression model $\mathbf{x}_i'\beta$, using a weight matrix $\mathbf{W}$.

Components of the vector of adjusted dependent variables z are assigned the values

$$z_i = \eta_i + \frac{(y_i - n_i p_i)}{n_i \, dp_i/d\eta_i},$$

where

- η_i is the current value of the linear predictor $\beta'\mathbf{x}_i$.
- y_i is the observed binomial count in the ith observation (taking values between $0, 1, \ldots, n_i$ for Bernoulli observations grouped by common covariate values).
- n_i is the number of binary observations in covariate group i.
- $\frac{dp_i}{d\eta_i}$ is the derivative of p_i with respect to the linear predictor η_i, evaluated at the current value of the linear predictor. For example, in a probit model, where $p_i = \Phi(\eta_i)$, $\frac{dp_i}{d\eta_i} = \phi(\eta)$, where $\Phi(\cdot)$ and $\phi(\cdot)$ represent the cumulative distribution and density functions of a standard normal random variable.

The weight matrix is defined to be the diagonal matrix $\mathbf{W}$ with elements

$$w_{ii} = \frac{n_i \left(\frac{dp_i}{d\eta_i}\right)^2}{p_i(1 - p_i)}$$

As in the case of the adjusted deviates, all variables are evaluated at their current estimates.

If we let $\mathbf{X}$ denote the matrix of explanatory variables with rows $\mathbf{x}_i'$ (including a column of 1's for the intercept, if present) and z the vector of adjusted dependent variates, then an iterative algorithm for obtaining the maximum likelihood estimates and asymptotic covariance matrix consists of the following steps:

0. Initialize the regression parameter $\hat{\beta}$ (a value of 0 is generally suitable).
1. Compute the linear predictor $\hat{\eta}_i = \mathbf{x}_i'\hat{\beta}$, or in matrix form $\hat{\eta} = \mathbf{X}\hat{\beta}$.
2. Compute the fitted probabilities $\hat{p}_i = F(\hat{\eta}_i)$.
3. Compute the derivative of the link function, $dp_i/d\eta_i = dF(\eta_i)/d\eta_i$.
4. Compute the adjusted covariate $\mathbf{z}$.
5. Compute the weight matrix $\mathbf{W}$.
6. Compute the asymptotic covariance matrix $\hat{\mathbf{C}} = (\mathbf{X}'\mathbf{W}\mathbf{X})^{-1}$.
7. Update the value of the regression coefficient according to $\hat{\beta} = \hat{\mathbf{C}}\mathbf{X}'\mathbf{W}\mathbf{z}$.

8. Repeat steps (1)–(7) until changes to the estimated regression coefficients $\hat{\beta}$ and the log-likelihood are sufficiently small. In most applications, six or seven iterations of the algorithm suffice.

At the termination of the algorithm, $\hat{\beta}$ contains the maximum likelihood estimate of the regression coefficients, $\hat{C}$ contains the asymptotic covariance matrix, and $\hat{p}$ contains the fitted probabilities.

3.9   Exercises

1. Hand et al. (1994) discuss data, stored in the file `swimmer.dat` (see Preface), from the 1990 Pilot Surf/Health Study of the New South Wales Water Board. Five variables were collected from a group of swimmers. The binary response variable EAR indicates the presence or absence of ear infections. There are four possible explanatory variables. The variable FREQ indicates if the swimmer is a frequent ocean swimmer (1 = yes, 0 = no), LOC indicates the swimming location (1 = beach, 0 = nonbeach), SEX is the gender (1 = male, 0 = female), and AGE is an age variable coded 2 if the age is between 15 and 19, 3 if the age is between 20 and 25, and 4 if the age is between 25 and 29. The purpose of this exercise is to determine which, if any, of the four variables are useful in predicting the occurrence of ear infections.

Let y_i denote a binary response variable equal to 1 if the ith swimmer has an ear infection and 0 otherwise. If p_i denotes the corresponding response probability, consider a logistic model of the form

$$\log\left(\frac{p_i}{1 - p_i}\right) = \beta_0 + \beta_1 \text{FREQ}_i + \beta_2 \text{LOC}_i + \beta_3 \text{SEX}_i + \beta_4 \text{AGE}_i,$$

where FREQ_i indicates if the ith swimmer is a frequent ocean swimmer, and so on.

a. Find maximum likelihood estimates and asymptotic standard errors of the regression coefficients of the logistic model.

b. As an alternative to the ML logistic model fit, run a MCMC algorithm to simulate from the posterior distribution of $\beta = (\beta_0, ..., \beta_4)'$, assuming a uniform prior for β. For each component of β, find the posterior mean and standard deviation and the 5th, 25th, 50th, 75th and 95th percentiles. Compare your analysis with the ML fit in part (a).

c. One way to assess the importance of each explanatory variable is to compute the posterior probability that the corresponding regression coefficient is positive, i.e., $\Pr(\beta_i > 0|\text{data})$. Compute this probability from the MCMC output for each covariate and decide which explanatory variables appear to be useful predictors.

d. Suppose that you believe a priori that beach swimmers do not run a greater risk of contracting ear infections than nonbeach swimmers. Specifically, you think that the probability of getting a ear infection is .2 if one is a beach swimmer, and the probability is the same .2 if one is a nonbeach swimmer. Suppose also that each guess is worth about 10 observations. Using a conditional means prior, summarize the posterior distribution of the regression vector β. Contrast this informative Bayesian analysis with the noninformative analysis done in part (b).

e. Find the deviance statistic for the full model including all four explanatory variables. Refit the ML logistic model four additional times, removing, in turn, each variable from the model. By use of a difference in deviances statistic, assess the importance of each variable in the model.

2. Ramsey and Schafer (1997) analyze data (contained in the file donner.dat) from an observational study by Grayson (1990) on the survival of the Donner and Reed families who traveled from Illinois to California by covered wagon in 1846. A number of the adult group did not survive the journey and it is of interest to see if the survival status is related to the gender and age of the individuals. For the ith adult in the party, denote by y_i the survival status y_i (1 if survive and 0 if did not survive) of the ith adult, and let AGE_i and SEX_i (1 if male and 0 if female) denote the corresponding age and sex variables. Let p_i denote the probability that the ith adult survives the journal. Assume that a logistic model of the form

$$\log \left(\frac{p_i}{1 - p_i} \right) = \beta_0 + \beta_1 AGE_i + \beta_2 SEX_i$$

describes the probability of a given adult's survival.

a. Find the ML estimate of the regression vector $\beta = (\beta_0, \beta_1, \beta_2)'$.
b. Check the goodness of fit of your ML fit by computing different sets of classical residuals. Inspect these residuals for possible outliers.
c. Assuming a vague prior, simulate a sample from the posterior distribution of β.
d. Graph the posterior distributions of the fitted probabilities p_i for all observations. Explain how the probability of survival depends on age and gender.
e. Compute the posterior distribution of the Bayesian residuals $\hat{p}_i - p_i$ and display them as a function of the age variable. Inspect this graph for possible outliers—contrast with the inspection of classical residuals in part (b).
f. Compute the cross-validation residuals and graph them as a function of AGE. Are the conclusions from this graph consistent with those in made in parts (b) and (e)?

3. Silvapulle (1981) discusses data which cross-classifies 120 individuals according to their gender, their score on a psychiatric screening questionnaire called the GHQ, and whether or not they were categorized as psychiatric "cases." For each possible score and gender, Table 3.10 provides the number of cases and noncases reported by gender and GHQ score. These data (stored in the file

| GHQ | Female | | Male | |
score	Case	Noncase	Case	Noncase
0	2	42	0	18
1	2	14	0	8
2	4	5	1	1
3	3	1	0	0
4	2	1	1	0
5	3	0	3	0
6	1	0	0	0
7	1	0	2	0
8	3	0	0	0
9	1	0	0	0
10	0	0	1	0

TABLE 3.10. Patient classifications obtained from GHQ questionnaire.

pcase.dat) are an example of grouped binomial data, where y_i can be regarded as the number of cases observed from a sample of size n_i, with explanatory variables $SCORE_i$ and $GENDER_i$.

a. Consider the logistic model of the form

$$\log\left(\frac{p_i}{1-p_i}\right) = \beta_0 + \beta_1 SCORE_i,$$

where p_i the probability that an individual with GHQ score $SCORE_i$ is considered to be a psychiatric case. Fit this model by maximum likelihood and by Bayesian methods using a noninformative prior on the regression vector. Compare the fits. Comment on the shape of the posterior distribution—is it well approximated by a normal distribution?

b. Fit the alternative probit model

$$\Phi(p_i)^{-1} = \beta_0 + \beta_1 SCORE_i$$

to the data using either a ML or Bayesian analysis. Compare the fits of the probabilities p_i from the logit and probit fits and comment on any differences you find.

c. Consider the inclusion of an additional term to the logistic model in (a) that incorporates gender. Using either a deviance test or a Bayesian test, is there evidence to suggest that gender is useful in predicting psychiatric cases when GHQ scores are already considered?

4. (Relationship between smoking and lung cancer.) In Chapter 2, data were collected to explore the relationship between smoking and lung cancer. Let y_i denote the binary response which indicates if the individual is a smoker or non-smoker and let p_i denote the probability that the individual is a smoker. In this example, one might suppose that the probability of smoking differs between control and cancer patients. This hypothesis can be represented by the logistic

model

$$\log\left(\frac{p_i}{1 - p_i}\right) = \beta_0 + \beta_1 s_i,$$

where s_i is the indicator variable which is equal to 1 if the corresponding individual is a cancer patient, and 0 if he is a control patient.

a. Using maximum likelihood, fit two models: the "full" model with all of the terms included, and a "reduced" model with the cancer/control variable removed. By the use of a deviance statistic, test the hypothesis that the patient type has no influence on the probability of smoking.

b. Construct an alternative test of the hypothesis that there is no relationship between cancer and smoking using a Bayes factor. Consider the use of the following two conditional means priors:

  • (Cancer/control effect should remain in the model.) The probability that a control patient smokes has a beta density with mean .9012 and, likewise, the probability that a cancer patient smokes has a beta distribution with the same mean (.9012). For each beta density, this information is worth 10 observations.

  • (Cancer/control effect should be removed from the model.) The probability that a control patient smokes has a beta density with mean .9012 and, likewise, the probability a cancer patient smokes has a beta distribution with the same mean (.9012). Assume this information is worth 10,000 observations.

Construct a Bayes factor in support of the hypothesis that the cancer/control effect should remain in the model. Contrast this value with the difference in deviances test in part (a).

5. In the National Longitudinal Survey of Youth, a random sample of American youth aged 15–22 in 1979 was surveyed each year from 1979 to 1992. In data from the 1992 survey, variables collected included the years of completed education (EDUC), the gender (MALE $= 1$ if male and 0 if female), the race (RACE $= 1$ if nonwhite and if white), the region of the United States of birth (EAST, MIDWEST, SOUTH, or WEST), and the number of years of completed schooling of the mother and the father. Since the number of school years of the mother and father are highly correlated, a new variable PEDUC was created which is equal to the average of the mother's and father's schooling. Suppose that one is interested in explaining the educational attainment based on the remaining four variables. To make the response variable binary, we define the variable BEDUC which is equal to 1 if the individual has completed at least one year of college and equal to 0 otherwise. Consider the model

$$\log\left(\frac{p_i}{1 - p_i}\right) = \beta_0 + \beta_1 \text{MALE}_i + \beta_2 \text{RACE}_i + \beta_3 \text{MIDWEST}_i$$
$$+ \beta_4 \text{SOUTH}_i + \beta_5 \text{WEST}_i + \beta_6 \text{PEDUC}_i,$$

where p_i denotes the probability that BEDUC $= 1$ for the ith individual. The data are contained in the file educ1.dat.

a. Fit the model using maximum likelihood. Find the estimates and the associated standard errors. Interpret each estimated regression coefficient in terms of the probability of completing at least one year of college. Specifically, find the fitted probability for individual A, who is male, white, lives in the Midwest, and whose parents are high school graduates (PEDUC = 12), and for individual B, who is female, nonwhite, lives in the South, and whose parents completed college (PEDUC = 16).

b. Generate a simulated sample from the posterior density of the regression vector. Find the posterior means and standard deviations of the regression coefficients. Find a 90% Bayesian interval estimate for the fitted probabilities of completing at least one year of college for the two individuals described in part (a).

c. Use at least two methods to assess how well the estimated probabilities agree with the binary outcomes, and comment on any particular observations that are not fit well by the model.

d. Using an analysis of deviance table to assess the importance of each covariate in the model. Find a new model which includes only the covariates that appear to significantly influence the probability of an individual completing at least one year of college.

6. In a survey of Duke University undergraduates, students were asked several questions, including:

Q1 How much did you learn in this course compared to all courses that you have taken at Duke?
1. Learned much less than average. 2. Learned less than average. 3. About average. 4. Learned more than average. 5. Learned much more than average.

Q2 How was this class graded?
1. Very leniently. 2. More leniently than average. 3. About average. 4. More severely than average. 5. Very severely.

Q3 What proportion of reading and written assignments did you complete?
1. <50%. 2. 50–75%. 3. 75–85%. 4. 85–95%. 5. >95%.

Q4 How does this instructor compare to all instructors that you have had at Duke?
1. One of the worst. 2. Below average. 3. Average. 4. Above average. 5. One of the best.

Responses to these survey questions are contained in the file survey.dat, available from our ftp site (see the Preface). The responses to these questions are provided in columns 20, 15, 7, and 27, respectively.

Create a binary response variable classifying a responses of 4 and 5 to the first question listed above as a success, and responses 1–3 as failures. Find a regression relation that predicts this binary response as some function of the responses of students to the remaining questions. Provide both the MLE and associated asymptotic standards errors, and the posterior means and posterior standard errors of all regression coefficients. Be sure to state and justify your prior assumptions on all model parameters.

7. (Continuation of Exercise 6.) For the final model that you selected in Exercise 6, compute the Pearson, deviance, and adjusted deviance residuals, and plot these residuals against fitted probabilities. Comment on any unusual observations. Next, compute the posterior and posterior-predictive residuals for this model. Plot the interquartile ranges of each of these residuals against their fitted values, and comment on this plot. Finally, using the same linear predictor used in the model you selected in Exercise 6, compute the posterior mean of the regression parameter for this model under a probit link, and plot the interquartile range of each latent residual distribution against fitted value. Construct a normal scores plot of the posterior mean of these latent residuals.

8. (Continuation of Exercise 6.) Provide an analysis of deviance table that includes the final model you selected in Exercise 6 and at least three other competing models. Using a normal approximation to the posterior in conjunction with Bayes theorem, calculate an approximate Bayes factor for each of these three competing models to the model that you selected. You may use either the approach outlined in Section 3.4.2 to specify priors on components of β, or you may use the (proper) priors that were used in Exercise 6 when computing these Bayes factors.

4

Regression Models for Ordinal Data

In the last chapter, we discussed regression models for binary data. These models were illustrated using grades of students in a statistics course. The grades themselves were not actually dichotomous—that is, they were not binary outcomes. Instead, we created dichotomous data by assigning a failing value of 0 to grades D and F, and a passing value of 1 to grades A, B, and C. By reducing the grades—which were originally recorded in five categories—to a dichotomous outcome, we threw away a significant portion of the information contained in the data. In this chapter, we explore techniques for analyzing this type of data without collapsing response categories by extending the binary regression models of Chapter 3 to the more general setting of ordered, categorical response data, or ordinal data.

Ordinal data are the most frequently encountered type of data in the social sciences. Survey data, in which respondents are asked to characterize their opinions on scales ranging from "strongly disagree" to "strongly agree," are a common example of such data. For our purposes, the defining property of ordinal data is that there exist a clear ordering of the response categories, but no underlying interval scale between them. For example, it is generally reasonable to assume an ordering of the form

strongly disagree < disagree < don't know < agree < strongly agree,

but it usually does not make sense to assign integer values to these categories. Thus, statements of the type

"disagree" − "strongly disagree" = "agree" − "don't know"

are not assumed.

4.1 Ordinal data via latent variables

The most natural way to view ordinal data is to postulate the existence of an under-lying latent (unobserved) variable associated with each response. Such variables are often assumed to be drawn from a continuous distribution centered on a mean value that varies from individual to individual. Often, this mean value is modeled as a linear function of the respondent's covariate vector.

This view of ordinal data was illustrated in Chapter 3 for the pass/fail response in the statistics class. In that example, we assumed that the logit of the pass probability could be expressed as a linear function of each student's SAT-M score. From a latent variable perspective, this model is equivalent to assuming that we can associate a latent performance variable with each student and that the distribution of this unobserved variable has a logistic distribution centered on a linear function of the student's SAT-M score. A geometric interpretation of this statement is provided in Figure 4.1. In this figure, a logistic distribution is centered on the linear predictor $x'\beta$, which in this instance is presumed to be 1. If the latent variable drawn from this distribution is greater than 0, then the student is assumed to pass the course. Otherwise, the student is assumed to fail.

The same geometric interpretation extends immediately to the original grade data if we introduce additional grade cutoffs. In the pass/fail version of this model, the point 0 represented the cutoff for a passing grade. As additional categorical

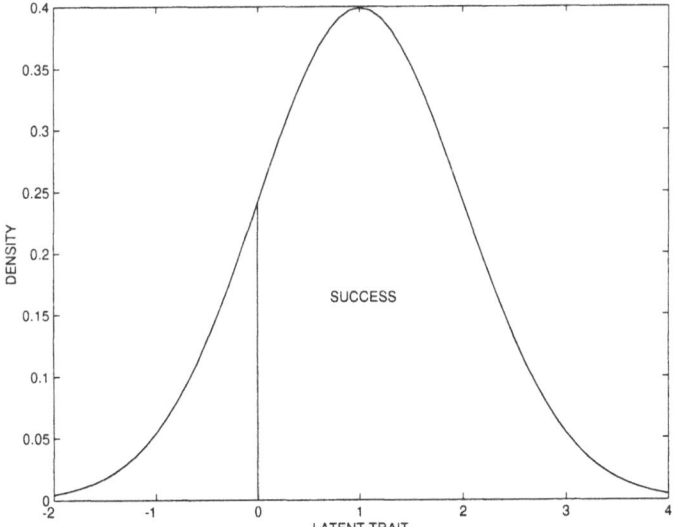

FIGURE 4.1. Latent trait interpretation of success probability. The logistic density repre-sents the distribution of latent traits for a particular individual. In modeling the event that this individual or item is a "success," we assume that a random variable is drawn from this density. If the random variable drawn falls below 0, a failure occurs; if it falls above 0, a success is recorded.

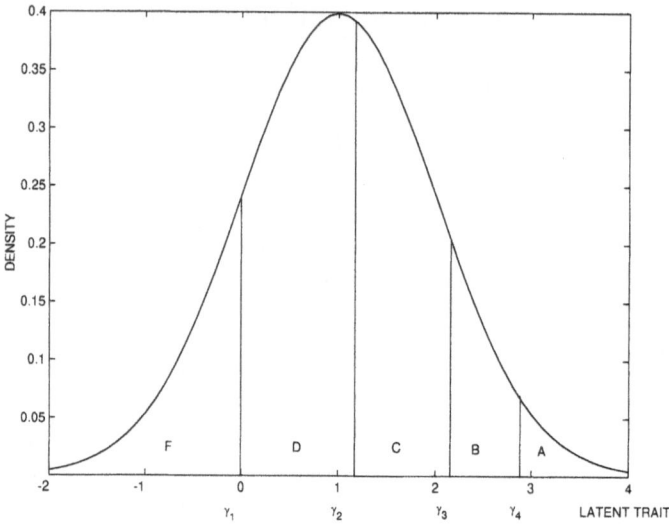

FIGURE 4.2. Latent trait interpretation of ordinal classification. In this plot, the logistic density again represents the distribution of latent traits for a particular individual. It is assumed that a random variable is drawn from this density, and the value of this random variable determines an individual's classification. For example, if a deviate of 0.5 is drawn, the individual receives a D.

responses or grade categories are introduced, we must create additional category or grade cutoffs within the model. For a class with five grades, a total of four additional grade cutoffs must be introduced. Also, because the response categories are ordered, we must impose a constraint on the values of grade cutoffs. Letting the upper grade cutoff for an F be denoted by γ_1, the upper grade cutoff for a D be denoted by γ_2, and so on, this ordering constraint may be stated mathematically as

$$-\infty < \gamma_1 \leq \gamma_2 \leq \gamma_3 \leq \gamma_4 \leq \gamma_5 \equiv \infty.$$

Note that the upper cutoff for an A, γ_5, is assumed to be unbounded. For notational convenience, we define $\gamma_0 = -\infty$.

Graphically, modeling the probability that the students in the statistics class received the grades A–F, rather than simply a pass or fail mark, requires only a minor modification of the situation depicted in Figure 4.1. Using the grade cutoffs $\gamma_1, \ldots, \gamma_4$, the expanded model is illustrated in Figure 4.2.

From Figure 4.2, we can imagine a latent variable Z that underlies the generation of the ordinal response. The distribution of this variable is governed by the equation

$$Z = \mathbf{x}'\beta + \epsilon, \tag{4.1}$$

where ϵ is a random variable drawn from a standard logistic distribution. When Z falls between the grade cutoffs γ_{c-1} and γ_c, the observation is classified into category c. To link this model for the data generation to the probability that an individual receives a particular grade, let f denote the density of the standard logistic distribution and let F denote the logistic distribution function. Denote by p_{ic} the probability that individual i receives a grade of c. Then, from (4.1), it follows that

$$
\begin{aligned}
p_{ic} &= \int_{\gamma_{c-1}}^{\gamma_c} f(z - x_i'\beta)dz \\
&= \Pr(\gamma_{c-1} < Z_i < \gamma_c) \\
&= F(\gamma_c - x_i'\beta) - F(\gamma_{c-1} - x_i'\beta).
\end{aligned}
\tag{4.2}
$$

The latent variable formulation of the problem thus provides a model for the probability that a student receives a particular grade in the course, or in the more general case, that a response is recorded in a particular category. If we further assume that the responses or grades for a sample of n individuals are independent of one another given these probabilities, the sampling distribution for the observed data is given by a multinomial distribution.

To specify this multinomial distribution, let us assume that there are C possible grades, denoted by $1, \ldots, C$. Also, suppose that n items are observed and that the grades or categories assigned to these n items are denoted by $y_1, \ldots y_n$; y_i denotes the grade observed for the ith individual.[1] Associated with the ith individual's response, we define a continuous latent variable Z_i and, as above, we assume that $Z_i = x_i'\beta + \epsilon_i$, where x_i is the vector of covariates associated with the ith individual and ϵ_i is distributed according to the distribution F. We observe the grade $y_i = c$ if the latent variable Z_i falls in the interval (γ_{c-1}, γ_c). Let p_i denote the vector of probabilities associated with assignment of the ith item into the categories $1, \ldots, C$; that is, $p_i = (p_{i1}, \ldots, p_{iC})$, where each element of p_{ic} denotes the probability that individual i is classified into category c. Let $y = (y_1, \ldots, y_n)$ denote the observed vector of responses for all individuals. It then follows that the probability of observing the data y, for a fixed value of the probability vectors $\{p_i\}$, is given by a multinomial density proportional to

$$
\Pr(y \mid p_i) \propto \prod_{i=1}^{n} p_{iy_i}.
\tag{4.3}
$$

---

[1] In defining the multinomial sampling density for an ordinal response, we assume that the multinomial denominator associated with each response is 1. For the more general case in which the ordinal responses are grouped by covariate, so that the multinomial denominator, say m_i, for the ith individual is greater than 1, this simply means that the m_i observations associated with the ith individual are considered independently in our model description. Because a multinomial observation with denominator greater than $m_i > 1$ can always be reexpressed as m_i multinomial observations with denominator 1, this distinction is irrelevant for most of the theoretical development discussed in this chapter and it somewhat simplifies notation and exposition. Of course, the likelihood function is unaffected by this change. The distinction only becomes important in defining the deviance statistic and individual deviance contributions; further comments concerning this point are delayed until Section 4.4.

Substituting the value of p_{ic} from (4.2) leads to the following expression for the likelihood function for β:

$$L(\beta, \gamma) = \prod_{i=1}^{n}[F(\gamma_{y_i} - \mathbf{x}_i'\beta) - F(\gamma_{y_i-1} - \mathbf{x}_i'\beta)]. \qquad (4.4)$$

If we parameterize the model by use of the latent data $\mathbf{Z}$ along with the parameters β and γ, then the likelihood, as a function of this entire set of unknown parameters and data, may be reexpressed in terms of the latent variables $\mathbf{Z}$ as

$$L(\beta, \gamma, \mathbf{Z}) = \prod_{i=1}^{n} f(Z_i - \mathbf{x}_i'\beta)I(\gamma_{y_i-1} \leq Z_i < \gamma_{y_i}), \qquad (4.5)$$

where $I(\cdot)$ indicates the indicator function and $\gamma = \{\gamma_1, \ldots, \gamma_C\}$ denotes the vector of grade cutoffs. Note that the latent variables Z_i may be integrated out of (4.5) to obtain (4.4).

4.1.1   Cumulative probabilities and model interpretation

Ordinal regression models are often specified in terms of cumulative probabilities rather than individual category probabilities. If we define

$$\theta_{ic} = p_{i1} + p_{i2} + \cdots + p_{ic}$$

to be the probability that the ith individual is placed in category c or below, then the regression component of an ordinal model of the form (4.2) may be rewritten

$$\theta_{ic} = F(\gamma_{ic} - \mathbf{x}_i'\beta). \qquad (4.6)$$

For example, if a logistic link function is assumed, Equation (4.6) becomes

$$\log\left(\frac{\theta_{ic}}{1 - \theta_{ic}}\right) = \gamma_c - \mathbf{x}_i'\beta. \qquad (4.7)$$

Note that the sign of the coefficient of the linear predictor is negative, as opposed to the positive sign of this term in the usual binary regression setting.

An interesting feature of model (4.7) is that the ratio of the odds for the event $y_1 \leq c$ to the odds of the event $y_2 \leq c$ is

$$\frac{\theta_{1c}/(1 - \theta_{1c})}{\theta_{2c}/(1 - \theta_{2c})} = \exp[-(\mathbf{x}_1 - \mathbf{x}_2)'\beta], \qquad (4.8)$$

independently of the category of response, c. For this reason, (4.7) is often called the *proportional odds model* (e.g., McCullagh, 1980).

Another common regression model for ordinal data, the proportional hazards model, may be obtained by assuming a complementary log-log link in (4.2). In this case,

$$\log[-\log(1 - \theta_{ic})] = \gamma_c - \mathbf{x}_i'\beta.$$

If one interprets $1 - \theta_{ic}$ as the probability of survival beyond (time) category c, this model may be considered a discrete version of the *proportional hazards model* proposed by Cox (1972). Further details concerning the connection between this model and the proportional hazards model may be found in McCullagh (1980).

Another link function often used to model cumulative probabilities of success is the standard normal distribution. With such a link, (4.2) becomes

$$\Phi^{-1}(\theta_{ic}) = \gamma_c - \mathbf{x}_i'\beta.$$

This model is referred to as the *ordinal probit model*. The ordinal probit model produces predicted probabilities similar to those obtained from the proportional odds model, just as predictions from a probit model for binary data produce predictions similar to those obtained using a logistic model. However, as we will see in Section 4.3.2, the ordinal probit model possesses a property that makes sampling from its posterior distribution particularly efficient. For that reason, it may be preferred to other model links (at least in preliminary studies) if a Bayesian analysis is entertained.

4.2   Parameter constraints and prior models

An ordinal regression model with C categories and $C-1$ unknown cutoff parameters $\gamma_1, ..., \gamma_{C-1}$ is overparameterized. To see this, note that if we add a constant to every cutoff value and subtract the same constant from the intercept in the regression function, the values of $\gamma_c - \mathbf{x}_i'\beta$ used to define the category probabilities are unchanged. There are two approaches that might be taken toward resolving this identifiability problem. The first is to simply fix the value of one cutoff, usually the first. This approach was implicitly taken in Chapter 3 when we considered binary regression and defined a success as an observation for which the latent value exceeded 0. In other words, we assumed that γ_1, the upper cutoff for a failure, was 0. A second approach that can be taken for establishing identifiability of parameters is to specify a proper prior distribution on the vector of category cutoffs, γ. For ordinal data containing more than three categories, a Bayesian approach toward inference requires that a prior distribution be specified for at least one category cutoff, regardless of which approach is taken. For that reason, we now turn our discussion to prior specifications for ordinal regression models.

4.2.1   Noninformative priors

In situations where little prior information is available, the simplest way to construct a prior distribution over the category cutoffs and regression parameter is to fix the value of one cutoff, usually γ_1, at 0.[2] After fixing the value of one cutoff, a uniform prior can then be assumed for the remaining cutoffs, subject to the

---

[2] When the constraint $\gamma_1 = 0$ is imposed, the values of the remaining cutoffs are defined relative to this constraint, and posterior variances of category cutoffs represent the variances of the contrasts $\gamma_c - \gamma_1$.

constraint that

$$\gamma_1 \leq \cdots \leq \gamma_{C-1}.$$

The components of the category cutoff vector and the regression parameter are assumed to be a priori independent, and a uniform prior is also taken for β.

This choice of prior results in a MAP estimate of the parameter values that is identical to the MLE. In general, these point estimators provide satisfactory estimates of the multinomial cell probabilities when moderate counts are observed in all C categories. However, if there are categories in which no counts are observed or in which the number of observations is small, the MLE and MAP estimates will differ significantly from the posterior mean. Furthermore, the bias and other properties of estimators of the extreme category cutoffs may differ substantially from the corresponding properties of estimators of the interior category cutoffs.

4.2.2   Informative priors

As in the case of binary regression, specifying a prior density on the components of β and γ can be difficult. A naive method for assigning an informative prior to these parameters proceeds by assuming that the vector β of regression parameters and the vector γ of cutpoints are independent and assign independent multivariate prior distributions to each. If the dimension of the regression vector is q and the dimension of the random component of the cutpoint vector γ is b, then a distribution of dimension $q + b$ is required for the specification of such prior. However, direct specification of the joint prior on (β, γ) is problematic because of the indirect effect that these parameters have on the multinomial probabilities of interest. Moreover, the task is made more difficult by the order restriction on the components of the category cutoffs.

A more attractive method of constructing an informative prior distribution generalizes the conditional means approach of Bedrick, Christensen, and Johnson (1996) that was used in the binary regression setting of Chapter 3. In setting the prior density using the conditional means approach, prior estimates of cumulative success probabilities are specified instead of specifying prior distributions on the values of the model parameters themselves. However, to establish identifiability of parameters in the conditional means prior, at least one cumulative probability of success must be specified for each model parameter. In other words, if there are four categories of response and $\gamma_1 = 0$, at least one prior guess must be made of the cumulative probability that a response is observed to be less than or equal to the second category ($y_i \leq 2$), and at least one prior guess must be made of the probability of observing at least one response less than or equal to the third category. In addition, the design matrix selected for the covariate values (including category cutoffs) must be nonsingular.

To illustrate the conditional means approach to specifying a prior, suppose that there are $C - 2$ unknown components of the cutoff vector γ (recall that $\gamma_1 = 0$ and $\gamma_C = \infty$) and q unknown components of the regression vector β. To construct a conditional means prior for $\{\gamma, \beta\}$, we must specify $M = q + C - 2$ values of the covariate vector x—call these covariate values $x_1, ..., x_M$. For each of the covariate vectors x_j, we specify a prior estimate and prior precision of our estimate of the corresponding cumulative probability $\theta_{(j)}$. Thus, for each covariate value, two items are specified:

1. A guess at the cumulative probability $\theta_{(j)}$ — call this guess g_j.
2. The prior precision of this guess in terms of the number of data equivalent "prior observations." Denote this prior sample size by K_j.

This prior information about $\theta_{(j)}$ can be incorporated into the model specification using a beta density with parameters $K_j g_j$ and $K_j(1 - g_j)$. If the prior distributions of the cumulative probabilities $\theta_{(1)}, ..., \theta_{(M)}$ are assumed to be independent, it follows that the joint prior density is given by the product

$$g(\theta_{(1)}, ..., \theta_{(M)}) \propto \prod_{j=1}^{M} \theta_{(j)}^{K_j g_j - 1} (1 - \theta_{(j)})^{K_j(1-g_j)-1}.$$

By transforming this prior on the cumulative probabilities back to (β, γ), the induced conditional means prior may be written

$$g(\beta, \gamma) \propto \prod_{j=1}^{M} \{ F(\gamma_{(j)} - x'_j \beta)^{K_j g_j} [1 - F(\gamma_{(j)} - x'_j \beta)]^{K_j(1-g_j)}$$
$$\times f(\gamma_{(j)} - x'_j \beta) \}, \tag{4.9}$$

subject to $\gamma_1 = 0 \leq \gamma_2 \leq \cdots \leq \gamma_{C-1}$. As before, $F(\cdot)$ denotes the link distribution function, and $f(\cdot)$ the link density.

4.3   Estimation strategies

Unlike the binary regression models of the previous chapter, maximum likelihood estimation routines for the ordinal regression models described above are often not supported in standard statistical packages. For this reason, it is useful to discuss iterative solution techniques for both maximum likelihood estimation as well as Markov chain Monte Carlo strategies for sampling from their posterior distributions of (γ, β).

4.3.1   Maximum likelihood estimation

Maximum likelihood estimates for ordinal regression models may be obtained using iteratively reweighted least squares (IRLS). An algorithm that implements IRLS can be found in the appendix at the end of this chapter. A byproduct of the algorithm is an estimate of the asymptotic covariance matrix of the MLE, which can be used to perform classical inference concerning the value of the MLE. If a uniform prior is assumed in a Bayesian analysis of the regression parameters, then the MLE and asymptotic covariance matrix obtained using IRLS also serve as approximations to the posterior mean and posterior covariance matrix.

4.3.2   MCMC sampling

In principle, the Metropolis-Hastings algorithm presented in Chapter 3 for sampling from the posterior distribution over a binary regression parameter can be adapted for sampling from the posterior distribution on the parameters in an ordinal regression model. There are, however, two important distinctions between the ordinal setting $(C > 2)$ and the binary setting.

First, multivariate normal proposal densities are not well suited for generating candidate vectors for ordinal regression parameters due to the ordering constraints imposed on the components of the category cutoff vector γ. Because of this ordering constraint, candidate vectors drawn from a multivariate normal density are rejected whenever the constraint on the components of γ is not satisfied. This can make generating candidate points inefficient, and for this reason, more complicated proposal densities are generally employed.

A second problem with standard Metropolis-Hasting schemes is that the probability that a candidate point is accepted often decreases dramatically as the number of classification categories increases. This effect is caused by the small probabilities that must invariably be associated with at least some of the observed categories and by the large relative change in the values of these probabilities assigned by candidate draws to the same categories. To overcome this difficulty, hybrid Metropolis-Hastings/Gibbs algorithms are often used to sample from the posterior distribution on ordinal regression parameters. Several such algorithms have been proposed for the case of ordinal probit models; among the more notable are those of Albert and Chib (1993), Cowles (1996), and Nandram and Chen (1996). Each of these algorithms exploits the latent data formulation described in Section 4.1. Here, we describe the Cowles' algorithm. It has the advantages that it is relatively simple to implement, displays good mixing, and extends to models with arbitrary constraints on the category cutoffs.

We describe the basic algorithm for ordinal probit models with the first cutoff parameter γ_1 fixed at 0 and uniform priors taken on (β, γ), where $\gamma = (\gamma_2, \ldots, \gamma_{C-1})$. Using the latent variable representation for the likelihood function (4.5), and letting ϕ denote the standard normal density, the joint posterior density of the latent variables and model parameters is given by

$$g(\beta, \gamma, \mathbf{Z} \mid \mathbf{y}) \propto \prod_{i=1}^{n} \phi(Z_i - \mathbf{x}_i' \beta) I(\gamma_{y_i-1} \leq Z_i < \gamma_{y_i}), \qquad (4.10)$$

for $-\infty < 0 < \gamma_2 < \cdots < \gamma_{C-1} < \infty$.

This representation of the joint posterior density suggests that a simple Gibbs sampling approach for simulating from the joint posterior of $(\mathbf{Z}, \beta, \gamma)$ might be possible because the full-conditional posterior distributions for $(\mathbf{Z}, \beta, \gamma)$,

- $g(\mathbf{Z} \mid \mathbf{y}, \beta, \gamma)$,
- $g(\beta \mid \mathbf{y}, \mathbf{Z}, \gamma)$,
- $g(\gamma \mid \mathbf{y}, \mathbf{Z}, \beta)$,

all have analytically tractable forms. The distribution of the components of $\mathbf{Z}$, given β and γ, have independent, truncated normal distributions where the truncation points are defined by current values of the category cutoffs. If we condition on $\mathbf{Z}$ and γ, the posterior density of the regression vector β is the same as the posterior density of the regression parameter for the standard normal model when the observational variance is known to be one. Finally, the conditional distribution of the components of the category cutoffs, for example γ_c, given current values of β and $\mathbf{Z}$, is uniformly distributed on the interval $(\max_{y_i=c-1} Z_i, \min_{y_i=c} Z_i)$. Unfortunately, when there are a large number of observations in adjacent categories, this interval tends to be very small, and movement of the components of γ will be minimal. This results in very slow mixing in Gibbs sampling schemes defined using these conditionals.

This difficulty can be resolved by means of an alternative simulation strategy for the category cutoffs. In Cowles' algorithm, this alternative strategy is obtained by partitioning the model parameters into two sets, $\{\mathbf{Z}, \gamma\}$ and β. Gibbs and MH sampling are then used to simulate in turn from the conditional distributions on the parameters in each set; that is, we alternately simulate between

- $g(\beta \mid \mathbf{y}, \mathbf{Z}, \gamma)$ and
- $g(\mathbf{Z}, \gamma \mid \mathbf{y}, \beta)$.

The conditional distribution of the regression parameter β is unchanged by this partitioning scheme and follows the multivariate normal distribution specified above. To simulate from the joint conditional posterior density of $(\mathbf{Z}, \gamma)$, we factor this density into the product

$$g(\mathbf{Z}, \gamma \mid \mathbf{y}, \beta) = g(\mathbf{Z} \mid \mathbf{y}, \gamma, \beta)g(\gamma \mid \mathbf{y}, \beta).$$

The composition method can then be applied to simulate γ from $g(\gamma \mid \mathbf{y}, \beta)$, and the components of $\mathbf{Z}$ can be sampled from $g(\mathbf{Z} \mid \mathbf{y}, \gamma, \beta)$. As noted in Section 4.1, $g(\gamma \mid \mathbf{y}, \beta)$ can be obtained analytically by integrating (4.10) over the latent variables to obtain

$$g(\gamma \mid \mathbf{y}, \beta) \propto \prod_{i=1}^{n}[\Phi(\gamma_{y_i} - \mathbf{x}_i'\beta) - \Phi(\gamma_{y_i-1} - \mathbf{x}_i'\beta)]. \qquad (4.11)$$

A Metropolis-Hastings step is used to sample from the conditional distribution of γ given $\mathbf{y}$ and β.

Steps in the hybrid Metropolis-Hastings/Gibbs sampler based on Cowles' algorithm can thus be summarized as follows.

Hybrid Metropolis-Hastings/Gibbs sampler:

0. Initialize $\beta^{(0)}$ and $\gamma^{(0)}$ (possibly to their MLE values); set $k = 1$ and $\sigma_{MH} = 0.05/C$. This value of σ_{MH} is simply a rule of thumb, and adjustments to σ_{MH} may be necessary if appropriate acceptance rates for γ are not obtained.
1. Generate a candidate $\mathbf{g}$ for updating $\gamma^{(k-1)}$:

   a. For $j = 2, ..., C - 1$, sample $g_j \sim N(\gamma_j^{(k-1)}, \sigma_{MH}^2)$ truncated to the interval $(g_{j-1}, \gamma_{j+1}^{(k-1)})$ (take $g_0 = -\infty$, $g_1 = 0$, and $g_C = \infty$).

b. Compute the acceptance ratio R according to

$$R = \prod_{i=1}^{n} \frac{\Phi(g_{y_i} - \mathbf{x}_i'\beta^{(k-1)}) - \Phi(g_{y_i-1} - \mathbf{x}_i'\beta^{(k-1)})}{\Phi(\gamma_{y_i}^{(k-1)} - \mathbf{x}_i'\beta^{(k-1)}) - \Phi(\gamma_{y_i-1}^{(k-1)} - \mathbf{x}_i'\beta^{(k-1)})} \tag{4.12}$$

$$\times \prod_{j=2}^{C-1} \frac{\Phi((\gamma_{j+1}^{(k-1)} - \gamma_j^{(k-1)})/\sigma_{MH}) - \Phi((g_{j-1} - \gamma_j^{(k-1)})/\sigma_{MH})}{\Phi((g_{j+1} - g_j)/\sigma_{MH}) - \Phi((\gamma_{j-1}^{(k-1)} - g_j)/\sigma_{MH})}.$$

c. Set $\gamma^{(k)} = \mathbf{g}$ with probability R. Otherwise, take $\gamma^{(k)} = \gamma^{(k-1)}$. Note that the second term in (4.12) accounts for the difference in the normalization of the proposal densities on the truncated normal intervals from which candidate points are drawn, whereas the first represents the contribution from the likelihood function. If a nonuniform prior is employed for γ, R should be multiplied by the corresponding ratio of the value of the prior at the candidate vector $(\beta^{(k-1)}, \mathbf{g})$ to the prior evaluated at $(\beta^{(k-1)}, \gamma^{(k-1)})$.

2. For $i = 1, \ldots, n$, sample $Z_i^{(k)}$ from its full conditional density given $\gamma^{(k)}$ and $\beta^{(k-1)}$. The full conditional density for $Z_i^{(k)}$ is a normal density with mean $\mathbf{x}_i'\beta^{(k-1)}$ and variance one, truncated to the interval $(\gamma_{y_i-1}^{(k)}, \gamma_{y_i}^{(k)})$.

3. Sample $\beta^{(k)}$ from a multivariate normal distribution given by

$$\beta^{(k)} \sim N((\mathbf{X}'\mathbf{X})^{-1}\mathbf{X}'\mathbf{Z}^{(k)}, (\mathbf{X}'\mathbf{X})^{-1})). \tag{4.13}$$

4. Increment $k = k + 1$ and repeat steps (1)–(3) until a sufficient number of sampled values have been obtained.

When implementing this algorithm, the acceptance rate for the category cutoffs should be monitored in Step 1. If this rate falls too much below 0.25 or above 0.5, σ_{MH} should be decreased or increased, respectively. Further details may be found in Cowles (1996).

A similar Metropolis-Hastings/Gibbs algorithm can be specified to accommodate non-Gaussian link functions—like the proportional odds and proportional hazards models—in a straightforward way. In step (1), the function $\Phi(\cdot)$ is replaced with the appropriate link distribution F. In step (2), the latent variables are sampled from the link distribution truncated to the interval bounded by the current estimates of the category cutoffs generated in step (1). For proportional hazards and proportional odds model, this is easily accomplished using the inversion sampling method (see Chapter 2). The most significant change to the algorithm is required in step (3). Without the normal link distribution, the full-conditional distribution of the regression parameter β will generally not be in a recognizable form. However, simple random-walk Metropolis-Hastings updates for β can be made using a multivariate normal proposal density centered on the previously sampled value of β and having covariance matrix $c(\mathbf{X}'\mathbf{X})^{-1}$. Values of the constant c in the interval $(0.5, 1)$ usually result in suitable acceptance rates.

4.4   Residual analysis and goodness of fit

Associated with every multinomial observation are C categories, and an individual's response (or absence of a response) in each of these categories can be used to define a residual. For binomial data (C=2), two such residuals are $y_i - n_i p_i$ and $(n_i - y_i) - n_i(1 - p_i)$. Of course, if you know the value of the first residual—that is, if you know p_i—you can figure out the value of the second, which depends only on $(1 - p_i)$ (since y_i and n_i are assumed known). The same is true for ordinal data with C categories; if you know the values of p_{ic} for $C - 1$ of the categories, you can figure out the probability for the last, since the probabilities have to sum to 1. Thus, for ordinal data, we potentially have $C - 1$ residuals for each multinomial observation.

This increase in dimensionality, from 1 to $C - 1$, complicates residual analyses. Not only are there more residual values to be examined, but the $C - 1$ residuals from each observation are correlated. It is therefore not clear how classical residuals (e.g., Pearson, deviance, and adjusted deviance residuals) should be displayed and analyzed. In the case of Bayesian residual analyses, the standard Bayesian residual and posterior-predictive residuals both involve $(C - 1)$-dimensional distributions, which, again, complicates model criticism. One possible solution to this problem is to create a sequence of binary residuals by collapsing response categories. For example, we might redefine a "success" as exceeding the first, second, ..., or $(C - 1)$st category. The resulting binary residuals can then be analyzed using the procedures described in Chapter 3, keeping in mind that the residuals defined for each success threshold are correlated. From a practical viewpoint, the binary residuals formed using exceedence of the extreme categories (categories 1 and $(C - 1)$) are often the most informative in identifying outliers, and so attention might be focused first on these residuals.

In contrast, residuals based on the vector of latent variables $\mathbf{Z}$ do not suffer from the problem of dimensionality, since only a single latent variable is defined for each individual. The latent residual for the ith observation is defined as before by

$$r_{i.L} = Z_i - x_i'\beta.$$

Nominally, the residuals $r_{1.L}, ..., r_{n.L}$ are independently distributed as draws from the distribution F. Deviations from the model structure are reflected by atypical values of these quantities from samples drawn from F. For this reason, case analyses based on latent residuals are generally easier to perform and interpret using scalar-valued latent residuals than using residuals defined directly in terms of observed data.

To judge the overall goodness of fit of an ordinal regression model, we can use the the deviance statistic, defined as

$$D = 2 \sum_{i=1}^{n} \sum_{j=1}^{C} I(y_i = j) \log(I(y_i = j)/\hat{p}_{ij}),$$

where $\hat{p}_{ij}$ denotes the maximum likelihood estimate of the cell probability p_{ij} and I is the indicator function. In this expression, the term $I(\cdot) \log(I(\cdot)/\hat{p}_{ij})$ is assumed to be 0 whenever the indicator function is 0. The degrees of freedom associated with the deviance statistic is $n - q - (C - 2)$, where q is the number of regression parameters in the model including the intercept. Asymptotically, the deviance statistic for ordinal regression models has a χ^2 distribution only when observations are grouped according to covariate values and the expected counts in each cell become large. When only one observation is observed at each covariate value, the deviance statistic is not well approximated by a χ^2 distribution.[3]

Besides its role as a goodness-of-fit statistic, the deviance statistic can be used for model selection. Perhaps surprisingly, the distribution of differences in deviance statistics for nested models is often remarkably close to a χ^2 random variable, even for data in which the expected cell counts are relatively small. The degrees of freedom of the χ^2 random variable that approximates the distribution of the difference in deviances is equal to the number of covariates deleted from the larger model to obtain the smaller model.

Related to the model deviance are the contributions to the deviance that accrue from individual observations. In the case of binary residuals, the signed square root of these terms were used to define the deviance residuals. However, for ordinal data, it is preferable to examine the values of the deviance contribution from individual observations directly, given by

$$d_i = 2 \sum_{j=1}^{C} I(y_i = j) \log(I(y_i = j)/\hat{p}_{ij}).$$

Observations that contribute disproportionately to the overall model deviance should be regarded with suspicion.[4]

---

[3] For grouped ordinal data, a more general definition of the deviance is needed. Letting y_{ij} denote the observed counts in category j for observation i, the deviance statistic can be defined more generally as

$$2 \sum_{i=1}^{n} \sum_{j=1}^{C} y_{ij} \log(y_{ij}/\hat{y}_{ij}),$$

where $\hat{y}_{ij}$ denotes the maximum likelihood estimate of the expected cell counts y_{ij}. As the expected number of counts in each cell of every observation approaches infinity (i.e., > 3), the distribution of this form of the deviance statistic approaches a χ^2 distribution, and so should be used for assessing goodness of fit whenever it is possible to group observations.

[4] For grouped ordinal data, an alternative definition of the deviance contribution from an individual observation is

$$\frac{2}{m_i} \sum_{j=1}^{C} y_{ij} \log(y_{ij}/\hat{y}_{ij}),$$

where $m_i = \sum_{j=1}^{C} y_{ij}$.

Turning to Bayesian case analyses, posterior-predictive residuals provide a generally applicable tool by which model adequacy can be judged and outlying observations can be identified.

Posterior-predictive residuals for ordinal data models are defined using the standard prescription discussed for binary models in Section 3.4, and samples from the posterior-predictive distribution of each observation can be obtained from an existing MCMC sample sequence of β and γ values, denoted by $\{\beta^{(j)}, \gamma^{(j)}, j = 1, \ldots, m\}$, using the following procedure:

For $j = 1, \ldots, m$ (m the MCMC run length):

1. For $i = 1, \ldots, C$, compute $\theta_{ic}^{(j)} = F(\gamma_{ic}^{(j)} - \mathbf{x}_i'\beta^{(j)})$ and set $p_{ic}^{(j)} = \theta_{ic}^{(j)} - \theta_{i-1,c}^{(j)}$.
2. Simulate observations y_i^* from multinomial distributions with success probabilities $p_{i1}^{(j)}, \ldots, p_{iC}^{(j)}$.

The posterior-predictive residual distribution can then be approximated using sampled values of the quantities

$$r_{i,PP} = y_i - y_i^*.$$

As in the case of binary regression, observations for which the residual posterior-predictive distributions are concentrated away from zero represent possible outliers.

4.5 Examples

4.5.1 Grades in a statistics class

We return now to the analysis of grades reported in Chapter 3. For convenience, the data from this example are reproduced in Table 4.1. We begin by illustrating maximum likelihood estimation for a proportional odds model. After discussing classical model checking procedures, we then discuss Bayesian analyses using both informative and noninformative priors.

Maximum likelihood analysis

As a first step in the analysis, we assume that the logit of the probability that a student receives an ordered grade of c or worse is a linear function of his or her SAT-M score; that is, we assume a proportional odds model of the form

$$\log\left(\frac{\theta_{ic}}{1 - \theta_{ic}}\right) = \gamma_c - \beta_0 - \beta_1 \times \text{SAT-M}_i. \tag{4.14}$$

Because an intercept is included in this relation, to establish identifiability we fix $\gamma_1 = 0$.

The maximum likelihood estimates and associated standard errors for the parameters γ and β are displayed in Table 4.2. The corresponding estimates of the

Student #	Grade	SAT-M score	Grade in previous statistics course
1	D	525	B
2	D	533	C
3	B	545	B
4	D	582	A
5	C	581	C
6	B	576	D
7	C	572	B
8	A	609	A
9	C	559	C
10	C	543	D
11	B	576	B
12	B	525	A
13	C	574	F
14	C	582	D
15	B	574	C
16	D	471	B
17	B	595	B
18	D	557	C
19	F	557	A
20	B	584	A
21	A	599	B
22	D	517	C
23	A	649	A
24	B	584	C
25	F	463	D
26	C	591	B
27	D	488	C
28	B	563	B
29	B	553	B
30	A	549	A

TABLE 4.1. Grades for a class of statistics students. The first column is student number. The second column lists the grade received in the class by the student, and the third and fourth columns provide the SAT-math score and grade for a prerequisite statistics course.

fitted probabilities that a student receives each of the five possible grades are plotted as a function of SAT-M score in Figure 4.3. In this figure, the white area reflects the probability that a student with a given SAT-M received an A, the lightly shaded area the probability of an B, and so on. From the plot, we see that the probability that a student with a 460 SAT-M score receives a D or F is about 57%, that a student scoring 560 on the SAT-M has approximately a 50% chance of receiving a B, and that a student who scored 660 on their SAT-M has a better than 80% chance of earning an A in the course.

It is interesting to compare the fit of the proportional odds model to the fit of the logistic regression model of Chapter 3, which was obtained by arbitrarily defining a passing mark as a grade of C or better. The maximum likelihood estimate of

Parameter	Estimate	Standard error
γ_2	2.22	0.64
γ_3	3.65	0.78
γ_4	6.51	1.33
β_0	−26.58	6.98
β_1	.0430	0.012

TABLE 4.2. Maximum likelihood estimates and standard errors for proportional odds model for statistics class grades example.

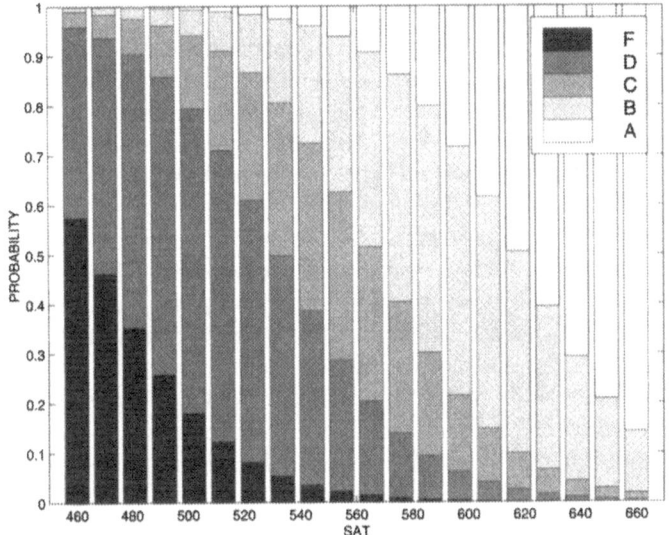

FIGURE 4.3. Fitted multinomial probabilities from the maximum likelihood fit of the proportional odds model. For each SAT value, the five shaded areas of the stacked bar chart represent the fitted probabilities of the five grades.

the probability that a student received a grade of C or higher under the logistic regression model was

$$\log\left(\frac{\hat{p}_i}{1 - \hat{p}_i}\right) = -31.11 + 0.058 \times \text{SAT-M}_i.$$

In the proportional odds model, θ_{i2} represents the probability that a student earns a grade of D or lower, so, $1 - \theta_{i2}$ is the probability that a student gets a C or higher. Under the proportional odds model, the latter probability may be expressed as

$$\log\left(\frac{1 - \theta_{i2}}{\theta_{i2}}\right) = -\gamma_2 + \beta_0 + \beta_1 \times \text{SAT-M}_i.$$

Using the estimates in Table 4.2, we see that the maximum likelihood estimate of this logit is

$$\log\left(\frac{1 - \hat{\theta}_{i2}}{\hat{\theta}_{i2}}\right) = -2.22 - 26.58 + 0.043 \times \text{SAT-M}_i.$$

This value is similar to that obtained using the binary regression model. The logit of the probability that a student received a C or better under the logistic model was estimated to increase at the rate of 0.058 units per SAT-M point; under the proportional odds model, this rate is 0.043. The asymptotic standard deviations of these slope estimates are 0.026 and 0.012, respectively.

The similarity of the parameter estimates obtained under the proportional odds model and the logistic model illustrates an important aspect of the ordinal modeling approach taken in this book. Due to the latent variable approach that underlies the model for both ordinal and binary responses, the interpretation of regression parameters is invariant with respect to the number of classification categories used. In the present case, the regression parameter β in the proportional odds model has the same interpretation as the regression parameter appearing in the logistic model, despite the fact that the grade categories A–C and D–F were collapsed to obtain the logistic model for binary responses. Of course, collapsing categories in this way results in some loss of information. This fact is reflected in the larger asymptotic standard deviations reported for the logistic model.

As a cursory check for model fit, we plotted the contributions to the deviance from individual observations against observation number in Figure 4.4. Interestingly, the most extreme deviant observation in the proportional odds fit appears to be student 19, rather than student 4 as it was in the logistic model. The reason for this difference becomes clear upon examining the data; student 19 received an F, while student 4 received only a D. Under the logistic model, both students were classified as failures, although student 19 apparently did much worse than student 4. Neither student's poor performance is well predicted by their SAT-M scores. It is also interesting to note that student 30's grade resulted in the second highest deviance contribution; this student had a slightly below average SAT-M score but received an A in the course. The grade of this student was not regarded as extreme in the logistic model.

For purposes of comparison, we next fit the ordinal probit model to the same data. In this case, the ordinal probit model takes the form

$$\theta_{ic} = \Phi\left(\gamma_c - \beta_0 - \beta_1 \times \text{SAT-M}_i\right). \tag{4.15}$$

As before, an intercept was included in this model since γ_1 was assigned the value 0. The maximum likelihood estimates for the probit model appear in Table 4.3.

As in the proportional odds model, one can plot the deviance contributions from each observation. The appearance of this plot was almost identical to Figure 4.4; thus, comments regarding the fit of the proportional odds model to individual student marks apply to the ordinal probit model as well. The similarity of the two deviance plots is a consequence of the fact that the fitted values under each

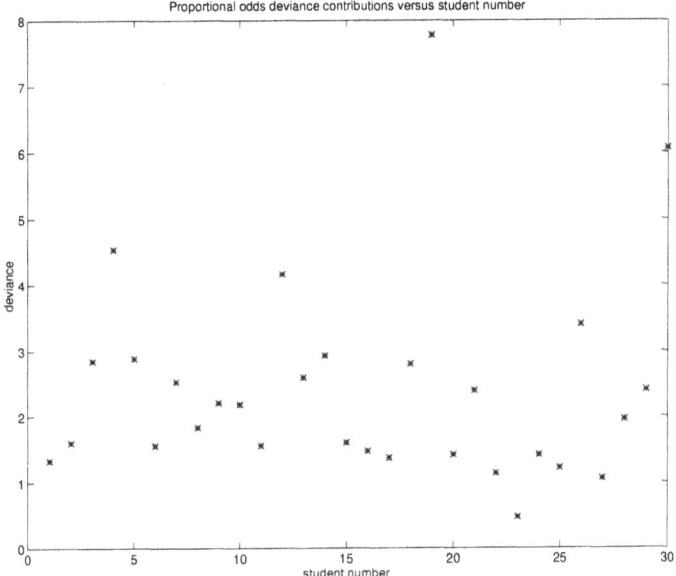

FIGURE 4.4. Deviance contributions in the proportional odds model for student grades. This plot does NOT depict deviance residuals. The square root of the deviance contributions was not taken, nor was there a natural way to attribute a sign to each observation.

Parameter	Estimate	Asy. std. dev.
γ_2	1.29	0.35
γ_3	2.11	0.41
γ_4	3.56	0.63
β_0	-14.78	3.64
β_1	0.0238	0.0063

TABLE 4.3. Maximum likelihood estimates and standard errors for ordinal probit model.

model are nearly identical. This point is illustrated in Figure 4.5, in which the predicted cell probabilities under the two models are plotted against one another. The deviance under the ordinal probit model was 73.5, while it was 72.7 under the proportional odds model.

Bayesian analysis with a noninformative prior

To further investigate the relationship between the student grades and SAT-M score, we next consider a Bayesian model using a vague prior on the parameters γ and β. Because of the similarity of fitted values obtained under the ordinal probit and proportional hazards model and the computational simplicity of sampling from the ordinal probit model using Cowles' algorithm, we restrict attention to the probit link.

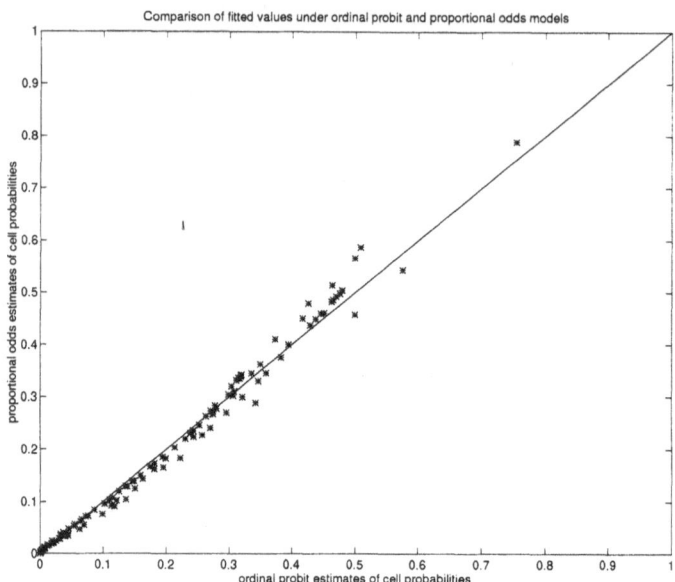

FIGURE 4.5. Fitted probabilities under ordinal probit model versus fitted probabilities for proportional odds model. All 150 predicted cell probabilities from the 30 observations are shown.

In applying Cowles' algorithm to these data, we initialized the parameter vectors with the maximum likelihood values. We then performed 20,000 MCMC iterations. The MCMC sample estimates of the posterior means of the parameter values are displayed in Table 4.4 and indicate that the posterior means agree well with the maximum likelihood estimates provided in Table 4.3. This fact suggests that the posterior distribution of the parameter estimates is approximately normal. The histogram estimates of the marginal posterior distributions displayed in Figure 4.7 support this conclusion.

A byproduct of the MCMC algorithm used to estimate the posterior means of the parameter estimates is the vector of latent variables $\mathbf{Z}$. As discussed at the end of Section 4.4, these variables provide a convenient diagnostic for detecting outliers and assessing goodness of fit. A priori, the latent residuals $Z_1 - x_1'\beta$, ..., $Z_n - x_n'\beta$ are a random sample from a $N(0, 1)$ distribution. Thus, deviations in the values of the latent residuals from an independent sample of standard normal deviates are symptomatic of violations of model assumptions.

A normal scores plot of the posterior means of the latent residuals are depicted in Figure 4.6. There are three points that appear to fall off of the 45° line. Recall from Section 3.4 that, since the posterior means are computed by averaging the sorted latent residuals across all iterations of the simulation, the points on the graph generally do not correspond to specific observations. However, from inspection of the latent residuals from the individual iterations, the smallest latent residual did correspond to observation 19 for 91% of the iterations, the next smallest residual

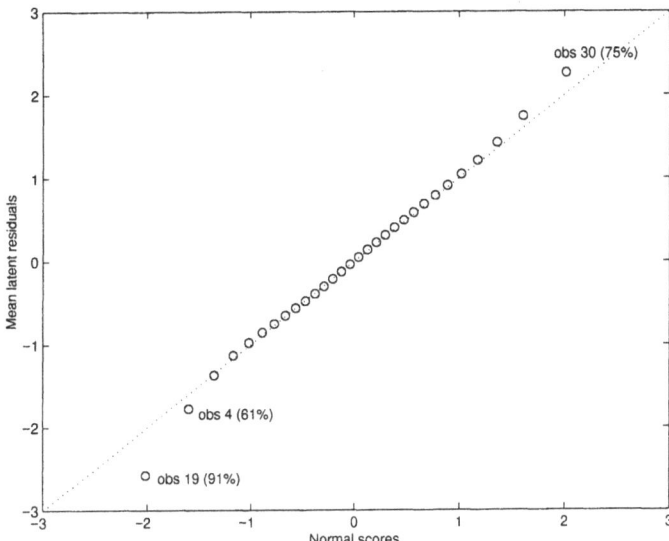

FIGURE 4.6. Normal scores plot of the posterior means of the sorted latent residuals from grades example. The labeled points indicate the percentages that particular observations contributed in the computation of the posterior mean residual.

Parameter	Post. mean	Post. std. dev.
γ_2	1.38	0.37
γ_3	2.26	0.42
γ_4	3.86	0.63
β_0	-12.05	3.73
β_1	.0257	0.0065

TABLE 4.4. Simulation estimates of the posterior means and standard errors for ordinal probit model using vague priors.

corresponded to observation 4 for 61% of the iterations, and observation 30 was the largest latent residual for 75% of the iterations. Thus, the most extreme latent residual posterior means appear to correspond to students 4, 19, and 30. With the exception of these three points, the normal scores plot does not suggest serious violations of model assumptions.

Bayesian analysis with an informative prior

We now illustrate how the methodology described in Section 4.2.2 can be used to specify a prior distribution for the parameters of the ordinal probit model for the statistics grades. Recall that in Chapter 3, we summarized our prior belief regarding the value of the regression parameter β through two prior estimates of the probability that a student received a grade of C or higher for two specified values of SAT-M. These estimates were that a student with a 500 SAT-M score

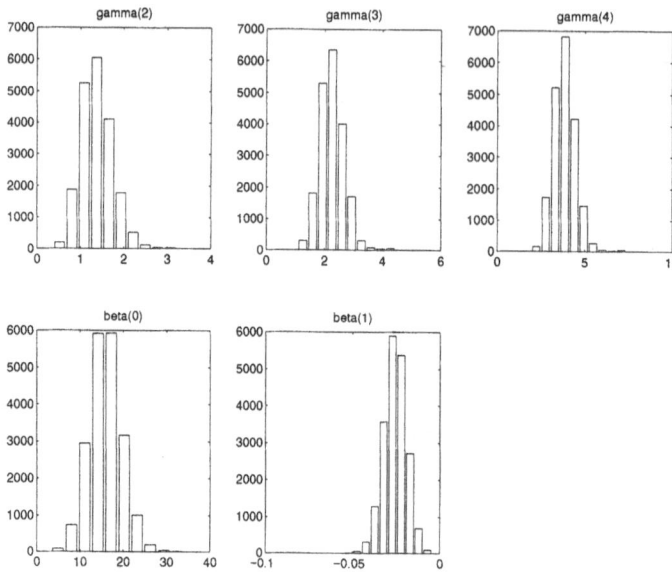

FIGURE 4.7. Histogram estimates of the marginal posterior distributions of the regression and category cutoff parameters in the statistics grades example.

would "pass" with probability 0.3 and that a student with an 600 SAT-M score would pass with probability 0.7. We also assigned a prior precision equivalent to five observations to each of these guesses. Here, we assume less certainty, and assign each guess the weight of one observation.

To specify a proper prior for all parameters in the ordinal probit model, we need to specify three more prior estimates for each of the three additional model parameters, γ_2, γ_3, and γ_4. Our three additional prior estimates are that the probability that a student having a 520 SAT-M score receives an F is 0.2, that the probability that a student with a 540 SAT-M score receives a C or lower is 0.75, and that the probability that a student with a 570 SAT-M score receives a B or lower is 0.85. As discussed above, we assign one observation's worth of information to each guess. The conditional means prior that results from these guesses can be expressed as follows:

$$
\begin{aligned}
g(\beta, \gamma) \propto \ & \Phi(-\beta_0 - 520\beta_1)^{0.2}(1 - \Phi(-\beta_0 - 520\beta_1))^{0.8} \\
& \times \Phi(\gamma_2 - \beta_0 - 500\beta_1)^{0.7}(1 - \Phi(\gamma_2 - \beta_0 - 500\beta_1))^{0.3} \\
& \times \Phi(\gamma_3 - \beta_0 - 540\beta_1)^{0.75}(1 - \Phi(\gamma_3 - \beta_0 - 540\beta_1))^{0.25} \\
& \times \Phi(\gamma_4 - \beta_0 - 570\beta_1)^{0.85}(1 - \Phi(\gamma_4 - \beta_0 - 570\beta_1))^{0.15} \\
& \times \Phi(\gamma_2 - \beta_0 - 600\beta_1)^{0.3}(1 - \Phi(\gamma_2 - \beta_0 - 600\beta_1))^{0.7} \\
& \times \phi(-\beta_0 - 520\beta_1)\phi(\gamma_2 - \beta_0 - 500\beta_1)\phi(\gamma_3 - \beta_0 - 540\beta_1) \\
& \times \phi(\gamma_4 - \beta_0 - 570\beta_1)\phi(\gamma_2 - \beta_0 - 600\beta_1).
\end{aligned}
\tag{4.16}
$$

Like the analysis with a uniform prior, the posterior density that results from this prior specification is not amenable to closed-form analysis. Thus, we must again resort to MCMC methods to obtain samples from the joint posterior distribution.

The MCMC algorithm described in Section 4.3 assumed a uniform prior for (β, γ). Modifying this algorithm for application with generic priors on γ and β requires the superposition of a Metropolis-Hastings step on the Gibbs sampler in step (3), and modification of the acceptance ratio in step (1). Letting

$$ s = \frac{g(\beta^{new}, \gamma^{new})}{g(\beta^{old}, \gamma^{old})} $$

denote the ratio of the prior density at the new parameter value to the old within each updating step, the required changes to Cowles' algorithm are

1. In step (1), the ratio R is multiplied by s. In this case, $\beta^{new} = \beta^{old} = \beta^{(k-1)}$, $\gamma^{old} = \gamma^{(k-1)}$, and $\gamma^{new} = \mathbf{g}$.
2. In step (3), take $\beta^{new} = \beta^{(k)}$, and $\gamma^{new} = \gamma^{old} = \gamma^{(k)}$. With probability equal to $\min(1, s)$, accept β^{new} as the new value of $\beta^{(k)}$. Otherwise, set $\beta^{(k)} = \beta^{(k-1)}$.

Posterior means and standard deviations estimated from a run of 10,000 iterates of this algorithm using the prior in (4.16) are provided in Table 4.5. Because the prior density used in this example was afforded six "prior observations," which is approximately one-fifth the weight of the data, the parameter estimates depicted in Table 4.5 differ noticeably from those obtained using a uniform prior.

To assess the effect of the informative prior information on the fitted probabilities for this model, in Table 4.6 we list the probabilities of various events using both noninformative and informative priors. The first row of this table provides the estimated probabilities that a student with a SAT-M score of 500 received a grade of D or lower in the statistics course under both informative and noninformative models. Note that the prior probability assigned to this event was 0.7. The posterior probability of this event using the noninformative prior analysis is 0.719; the corresponding posterior probability using the informative prior is 0.568. The prior information lowers this probability *below* its expected prior value, although this apparently nonlinear effect can be explained by examining the other probabilities in Table 4.6. In this table, we see that the overall effect of the informative prior is to shift the estimated posterior probabilities obtained under the noninformative posterior toward the prior estimates.

Parameter	Post. mean	Post. std. dev.
γ_2	1.09	0.28
γ_3	1.80	0.34
γ_4	2.85	0.44
β_0	-5.68	2.73
β_1	.0132	0.0048

TABLE 4.5. Simulation estimates of the posterior means and standard errors for ordinal probit model using using an informative prior.

Event	Prior prob.	Noninf. post.	Inf. post.
(500, D or lower)	.7	.719	.568
(520, F)	.2	.094	.118
(540, C or lower)	.75	.667	.638
(570, B or lower)	.85	.896	.843
(600, D or lower)	.3	.023	.125

TABLE 4.6. Prior and posterior probability estimates of particular events using noninformative and informative priors. The notation (540, D or lower) refers to the event that a student with a SAT-M score of 540 receives a grade of D or lower.

4.6   Prediction of essay scores from grammar attributes

A problem faced by large educational testing companies (e.g., ETS, ACT) involves grading thousands of student essays. As a result, there is great interest in automating the grading of student essays or—failing this—determining easily measurable qualities of essays that are associated with their ranking. The purpose of this example is to study the relationships between essay grades and essay attributes. The data in this example consist of grades assigned to 198 essays by 5 experts, each of whom rated all essays on a 10-point scale. A score of 10 indicates an excellent essay. Similar data have also been analyzed by, for example, Page (1994) and Johnson (1996). For present purposes, we examine only the grades assigned by the first expert grader, and the essay characteristics of average word and sentence length, number of words, and the number of prepositions, commas, and spelling errors.

Following a preliminary graphical analysis of the data, we chose to examine the predictive relationships between an expert's grade of an essay and the variables square root of the number of words in the essay (SqW), average word length (WL), percentage of prepositions (PP), number of commas × 100 over number of words in the essay (PC), the percentage of spelling errors (PS), and the average sentence length (SL). Plots of each of these variables versus the essay grades are displayed in Figure 4.8.

Based on the plots in Figure 4.8, we posited a baseline model of the form

$$\Phi^{-1}(\theta_{ic}) = \gamma_c - \beta_0 - \beta_1 WL - \beta_2 SqW - \beta_3 PC - \beta_4 PS - \beta_5 PP - \beta_6 SL, \quad (4.17)$$

where, as before, θ_{ic} denotes the cumulative probability that the ith essay received a score of c or below and Φ denotes the standard normal distribution function. The maximum likelihood estimates for this model are displayed in Table 4.7.

The deviance of model (4.17) was 748.7 on $198 - 15 = 183$ degrees of freedom, using the usual convention that the number of degrees of freedom in a generalized linear model is equal to the number of observations less the number of estimated parameters. The deviance statistic is much larger than the degrees of freedom, suggesting some overdispersion in the model. This confirms our prior intuition that the six explanatory variables in the model cannot accurately predict the grades assigned by any particular human expert. (In fact, we might expect considerable variation between the grades assigned by different experts to the same essay.)

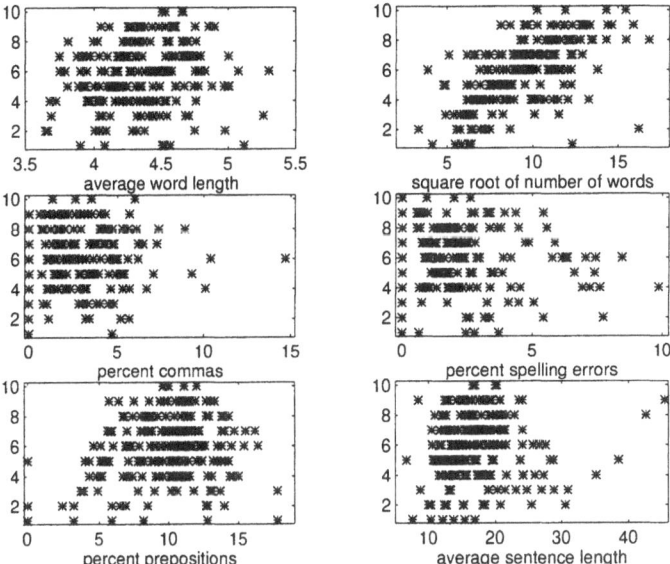

FIGURE 4.8. Plots of essay grades obtained from the first expert grader versus six explanatory variables.

Thus, to interpret the standard errors in the table, it is probably prudent to apply a correction for overdispersion. Because the usual estimate of overdispersion for ordinal regression models is deviance/degrees of freedom, in this case 4.09, each of the standard errors in Table 4.7 should be multiplied by the square root of the estimated overdispersion (≈ 2.0) to obtain a more realistic estimate of the sampling uncertainty associated with each parameter.

To investigate the source of overdispersion, it is interesting to examine the deviance contribution from each essay grade. To this end, a plot of deviance contribution versus the square root of the number of words is provided in Figure 4.9. As illustrated in the figure, the deviance of several observations exceeds 8, and the deviance for two observations exceeds 14. The values 8 and 14 correspond approximately to the 0.995 and 0.9998 points of a χ_1^2 random variable, although it is unlikely that the asymptotic distribution of either the total deviance or the deviance of individual observations is well approximated by a χ^2 random variable. However, the large values of the deviance associated with these observations provides further evidence that the grammatical variables included in the model do not capture all features used by the grader in evaluating the essays.

From Table 4.7 and the preliminary plots of the essay grades versus explanatory variables, it is clear that several of the variables included in the baseline model were not significant in predicting essay grade. To explore which of the variables should be retained in the regression function, we used a backward selection procedure in which variables were excluded sequentially from the model. The results of this procedure are summarized in the analysis of deviance table displayed in Table

Parameter	MLE	Asy. std. dev.
γ_2	0.632	0.18
γ_3	1.05	0.20
γ_4	1.63	0.21
γ_5	2.19	0.22
γ_6	2.71	0.23
γ_7	3.39	0.24
γ_8	3.96	0.26
γ_9	5.09	0.35
β_0	-3.74	1.08
β_1	0.656	0.23
β_2	0.296	0.032
β_3	0.0273	0.032
β_4	-0.0509	0.038
β_5	0.0461	0.023
β_6	0.00449	0.013

TABLE 4.7. Maximum likelihood estimates and asymptotic standard errors for the baseline regression model for essay grades.

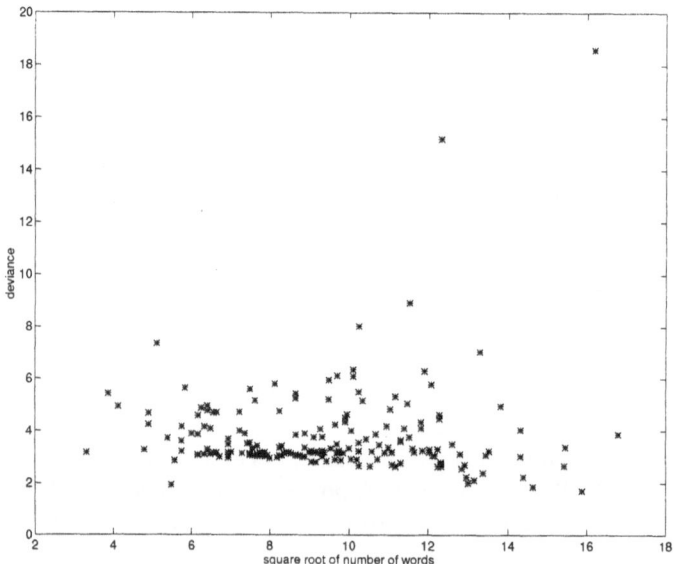

FIGURE 4.9. Deviance contribution from individual essay grades versus the square root of the number of words contained in each essay.

4.8. In this table, the entry on the third line (labeled "-PC") provides the reduction of deviance and adjusted deviance that result from deleting the variable PC from the probit model containing all covariates in the full model, excluding those already deleted, in this case SL. Both the reduction in deviance associated with

Model	Change in deviance	Corrected change in deviance
Full Model	—	—
-SL	0.12	0.03
-PC	0.70	0.17
-PS	1.77	0.43
-PP	3.98	0.97
-WL	7.90	1.93
-SqW	84.08	20.56

TABLE 4.8. AOD table for essay grades. The entries in the second column represent the increase in deviance resulting from deletion of the variable indicated in the first column, as compared to the model on the previous row. For example, the last row provides the difference in model deviance between a model containing only the category cutoffs and SqW and a model with covariates comprised of the category cutoffs, WL, and SqW. Entries in the third column represent the entries in the second columns divided by 4.09, the estimate of the model overdispersion from the full model.

the deletion of each model variable and the reduction in the deviance corrected for overdispersion are provided. As in Chapter 3, the overdispersion of the model was calculated as the model deviance divided by the degrees of freedom. Further motivation for this correction for overdispersion may be found in, for example, McCullagh and Nelder (1989).

By comparing the corrected changes in deviance displayed in the table to the corresponding tail probabilities of a χ_1^2 random variable, it appears that the variables SL, PC, and PS (average sentence length and percentage of commas and spelling errors) were not important in predicting the essay scores assigned by this grader. Likewise, the variable PP (percentage of prepositions) appears to be only marginally significant as a predictor, while the variables WL and SqW (word length and square root of number of words) are significant or highly significant. These results suggest that the variables SL, PC, and PS might be excluded from the model, leaving a predictive model of the form

$$\Phi^{-1}(\theta_{ic}) = \gamma_c - \beta_0 - \beta_1 \mathrm{WL} - \beta_2 \mathrm{SqW} - \beta_3 \mathrm{PP}. \qquad (4.18)$$

Turning next to a default Bayesian analysis of these data, if we assume a vague prior on all model parameters, we can use the MCMC algorithm described in Section 4.3.2 to sample from the posterior distribution on the parameters appearing in either the full model (4.17) or the reduced model (4.18). For purposes of illustration, we generated 5,000 iterates from the full model and used these sampled values to estimate the posterior means of the regression parameters. These estimates are provided in Table 4.9 and are quite similar to the maximum likelihood (and, in this case, maximum a posteriori) estimates listed in Table 4.7.

Bayesian case analyses based on output from the MCMC algorithm proceeds as in the previous example. By saving the latent variables values generated in the MCMC scheme, we can construct a normal scores plot of the posterior means of the latent residuals, as depicted in Figure 4.10. Like the deviance plot, this figure

Parameter	Post. mean	Post. std. dev.
γ_2	0.736	0.16
γ_3	1.19	0.18
γ_4	1.79	0.21
γ_5	2.35	0.21
γ_6	2.88	0.22
γ_7	3.59	0.22
γ_8	4.18	0.24
γ_9	5.30	0.30
β_0	−3.76	1.12
β_1	0.670	0.24
β_2	0.305	0.033
β_3	0.0297	0.033
β_4	−0.0520	0.038
β_5	0.0489	0.024
β_6	0.00463	0.013

TABLE 4.9. Posterior means of parameter estimates and standard errors for the full regression model for the essay grades.

suggests that the smallest two residuals are unusually small for this particular model. In the computation of the posterior means for these two smallest residuals, we see from Figure 4.10 that these posterior means were small primarily due to the contributions of observations 3 and 45. There is also evidence that the distribution of the latent residuals is non-Gaussian, due to the snake-like appearance of this graph.

In addition to the latent residuals, we can also examine the posterior-predictive residuals to investigate the overdispersion detected in the likelihood-based analysis. As in Chapter 3, if we let y_i^* denote a new essay grade based on the posterior-predictive distribution for covariate value $\mathbf{x}_i$ and we let y_i denote the observed grade of the ith essay, then the posterior-predictive residual distribution for the ith observation is defined as the distribution of $y_i - y_i^*$.

A plot of the estimated interquartile ranges of the posterior-predictive residuals is provided in Figure 4.11. The appearance of this plot is similar to Figure 3.12, and like Figure 3.12, it indicates model lack of fit. To more formally quantify this lack of fit, we might again posit a random effects model, but in this case there are at least two distinct sources of error which we would like to model. The first is the inability of the regression model to fully explain the nuances of human graders; the regression model clearly cannot account for all of the essay attributes used by the expert in arriving at a grade for an essay. The second is the variability between experts in assigning grades to essays. As we mentioned at the beginning of this example, there were four other experts who also assigned grades to these same essays, and there was considerable disagreement among the experts on the appropriate grade for any particular essay. Thus, a simple random effects model is unlikely to capture both sources of overdispersion, which suggests that more a comprehensive model is needed. We investigate such models in the next chapter.

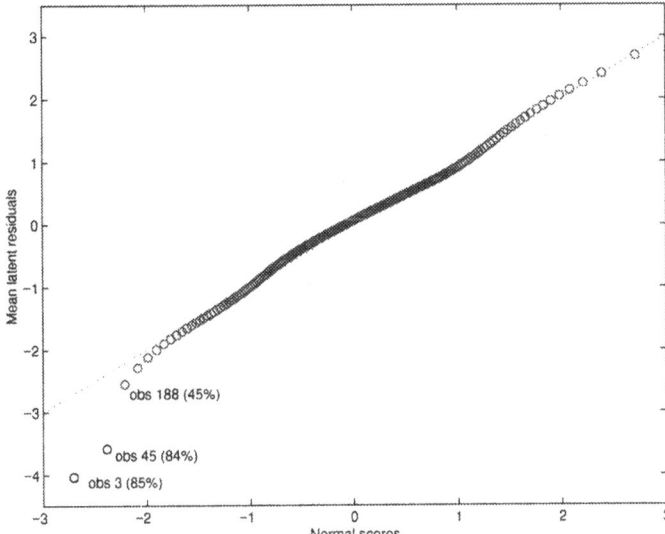

FIGURE 4.10. Normal scores plot of the posterior means of the sorted latent residuals for essay grading example. The labeled points indicate the percentages that particular observations contributed in the computation of the posterior mean residual.

4.7   Further reading

Classical ordinal regression models based on cumulative probabilities are described in McCullagh (1980) and Chapter 3 of Fahrmeir and Tutz (1994). Albert and Chib (1993), Cowles, Carlin, and Connett (1996), and Bradlow and Zaslavsky (1999) illustrate Bayesian fitting of ordinal regression models using latent variables.

4.8   Appendix: iteratively reweighted least squares

The following iteratively reweighted least squares (IRLS) algorithm can be used to find both the maximum likelihood estimate and asymptotic covariance matrix for parameters appearing in the ordinal regression models described in Chapter 4. Algorithmically, implementing IRLS requires definitions of a working dependent variable, a matrix of regressors, and regression weights at each update. To define these variables, begin by letting α denote the vector of model parameters, $(\gamma_2, \ldots, \gamma_C, \beta_0, \ldots, \beta_p)$. Note that γ_1 is not included in this vector because its value is assumed to be 0. For concreteness, assume that there are five response categories.

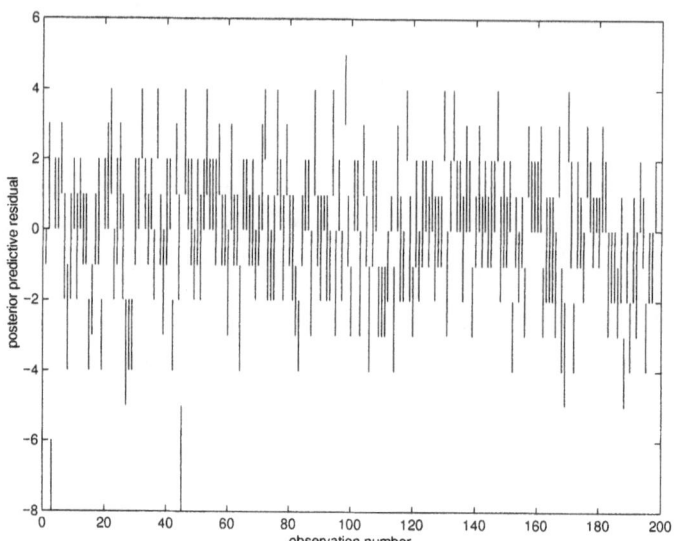

FIGURE 4.11. Interquartile ranges of the posterior predictive residuals. The fact that a high proportion of these ranges do not cover 0 is an indication of overdispersion, or other lack-of-fit.

Assuming five response categories, define

$$
X_i = \begin{bmatrix} 0 & 0 & 0 & -\mathbf{x}_i' \\ 1 & 0 & 0 & -\mathbf{x}_i' \\ 0 & 1 & 0 & -\mathbf{x}_i' \\ 0 & 0 & 1 & -\mathbf{x}_i' \end{bmatrix}
$$

where the covariate vectors $\mathbf{x}_i$ are preceded by a $(C-1) \times (C-1)$ identity matrix with the first column omitted (corresponding to γ_1, which is 0). Also, take

$$
S_i = \begin{bmatrix} 1 & 0 & 0 & 0 \\ -1 & 1 & 0 & 0 \\ 0 & -1 & 1 & 0 \\ 0 & 0 & -1 & 1 \\ 0 & 0 & 0 & -1 \end{bmatrix}
$$

to be a $C \times (C-1)$ array, and define $H_i = \mathrm{diag}(f_{i1}, \ldots, f_{i,C-1})$, where f_{ic} is the derivative of the link distribution evaluated at $\gamma_c - \mathbf{x}_i'\beta$. Finally, take $V_i = \mathrm{diag}(\mu_i)$, where $\mu_i = \mathbf{p}_i$, the vector of response probabilities for the ith observation.

With this admittedly tedious notation in place, the components of the IRLS algorithm can be defined as follows. Let the working dependent vector be $\mathbf{w} = (w_1', \ldots, w_n')'$, where

$$
w_i = S_i H_i X_i + (\mathbf{y}_i - \mu_i),
$$

and let the regression matrix be defined as

$$\mathbf{R} = (X_1' H_1 S_1', \ldots, X_n' H_n S_n')'.$$

Finally, let the weight matrix be $\mathbf{V} = \text{diag}(V_1^{-1}, \ldots, V_n^{-1})$.

To implement IRLS, each of the components of the algorithm must be initialized. The most reliable way to accomplish this initialization is to base the initial values on the observed counts y. For example, μ_i can be initialized by taking

$$\mu_{ic} = \frac{I(y_i = c) + 1/2}{1 + C/2}.$$

Similar initializations follow for other components of the algorithm.

After initialization, a least squares estimation of the linear equation

$$\mathbf{w} = \mathbf{R}\alpha$$

with weight matrix $\mathbf{V}$ is performed recursively until changes in the regression vector α are negligible. Note that $\mathbf{w}$, $\mathbf{R}$, and $\mathbf{V}$ must all be updated using the new value of α obtained after each least squares update.

Upon convergence of the algorithm, the information matrix is

$$\mathbf{I_F} = \sum_{i=1}^{n} X_i' H_i S_i' V_i^{-1} S_i H_i X_i,$$

and $-\mathbf{I_F}^{-1}$ represents the asymptotic covariance matrix of the MLE $\hat{\alpha}$.

Further details concerning this algorithm may be found in Jansen (1991).

4.9 Exercises

1. Reconsider the educational achievement data of Exercise 5 of Chapter 3. In the dataset educ.dat, the response variable "years of education" is coded into the five ordered categories, where the codes 1–5 denote the categories "not completed high school", "completed high school", "had some college", "completed college", and "some graduate education", respectively.

   a. Find the MLE and associate standard errors for the full model (probit link), which includes the covariates gender, race, region and parents' education. Constrast your fitted model with the estimated binary regression model found in Exercise 5 of Chapter 3.
   b. Refit the ordinal model using a maximum likelihood logistic fit. Construct a scatterplot of the fitted probabilities from the logistic and probit fits. Comment on any observed differences between the two fits.
   c. Use the MCMC algorithm described in Section 4.3 to sample from the posterior distribution of the ordinal regression model using a noninformative prior and a probit link. Compute the posterior means and standard deviations of the individual regression coefficients. Compare your answers with the MLE estimates and standard errors found in part (a).

d. Plot the deviance contributions from the MLE fit against the observation number to assess model fit. In addition, construct a normal probability plot of the posterior means of the sorted latent residuals from the Bayesian fit. Comment on any unusual observations that are not well explained by the fitted model.

2. Reanalyze the data of Exercise 6 of Chapter 3 without collapsing the categories of the response variable. Provide both the MLE and associated asymptotic standard errors, and the posterior means and posterior standard errors of all regression coefficients. Be sure to state and justify your prior assumptions on all model parameters. Compare the results of this analysis to the results obtained using the definition of the response specified in Exercise 6 of Chapter 3 (stored in the file survey.dat on web locations cited in Preface).

3. (Continuation of Exercise 2.) Fit the same linear predictor used in Exercise 2 to the grade data using a probit link. Using Cowles' algorithm to generate samples from the posterior distribution on the regression parameters in your model, plot histogram estimates of the marginal posterior distribution of each parameter.

4. (Continuation of Exercise 2.) For the final model that you selected in Exercise 2, plot the contribution to the deviance by observation against observation number. Next, compute the posterior-predictive residuals for this model. Plot the interquartile ranges of each of these residuals against their fitted values, and comment on this plot. Finally, using the same linear predictor used in the model you selected in Exercise 2, compute the posterior mean of the regression parameter for this model under a probit link, and plot the interquartile range of each latent residual distribution against fitted value. Construct a normal scores plot of the posterior mean of these latent residuals.

5. (Continuation of Exercise 2.) Provide an analysis of deviance table that includes the final model you selected in Exercise 2 and at least three other competing models. Using a normal approximation to the posterior in conjunction with Bayes theorem (see Section 2.3.1), calculate an approximate Bayes factor for each of these three competing models to the model that you selected. Be sure to specify the prior distributions used for each of the models. Compare model selection based on the AOD table to model selection using Bayes factors. For the model that you selected as "best," compute the fitted values of the response probabilities for several values of the covariates in your model. Comment. Provide an interpretation of your model coefficients and conclusions for a nonstatistician.

6. In a study of the development of Downs syndrome children, Skotko collected survey data from 55 parents concerning the extent and type of language training provided to children, along with pertinent developmental response variables. The following questions were scored on a six-point scale, in which the first category was "Not applicable," and the remaining questions ranged from "never" to "very frequently," in that order.

a. Were speech therapists/pathologists involved in your child's development before the age of 5?

b. Were speech therapists/pathologists involved in your child's development at the age of 5 or later?

c. Was sign language used with your child before the age of 5?

d. Was sign language used with your child at the age of 5 or later?

e. Were nutritional supplements used before the age of 5?

f. Were nutritional supplements used at the age of 5 or later?

g. Did you read books to your child before the age of 5?

h. Did you read books to your child at the age of 5 or later?

Interest in this study focused on the relationship between the answers to the questions above and several outcome variables, including parent response to the question "How well is your child able to maintain a conversation with a friend?" This question was scored on a 5-point scale, with categories ranging from "(1) Completely unable" to "(5) Extremely well." Treating the answer to this question as your response variable, investigate ordinal regression models that predict this response using the questions described in the preceding paragraph. Provide both the MLE and associated asymptotic standards errors, and the posterior means and posterior standard errors of all regression coefficients. Be sure to state and justify any prior assumptions you make for your model parameters and provide a substantive interpretation of your results. Data from the study is contained in the file skotko.dat obtained from the web site referenced in the Preface.

7. (Continuation of Exercise 6.) Conduct both classical and Bayesian case analyses for the final model you selected in the previous question. Comment on your results.

8. (Continuation of Exercise 6.) Aside from the "best" model that you selected in Exercise 6, identify at least three other competing models, at least one of which is not nested in your model, and display an AOD table for these models. Next, specify a prior distribution for the parameters in each of these models and provide approximate Bayes factors of each alternative model to the model you selected. Compare the AOD selection procedure to that based on Bayes factors.

5

Analyzing Data from Multiple Raters

At the end of the last chapter, we examined the relationship between an expert's ratings of a set of high school student's essays and several easily quantifiable attributes measured from these essays. The particular essay grades that we examined happened to be the grades from the first of five experts who graded the essays. However, with more than one expert grader, an obvious question becomes "How would our analysis change if we used another expert's ratings, or if we somehow combined the grades from all experts?"

Collecting ordinal data from multiple raters is a bit like having more than one wristwatch. With one watch, you know the time—or at least you think you do. But with two watches, you are never sure. The same phenomenon exists with multiple raters. In the previous chapter, we assumed that the "true grade" of each essay was known, and then we analyzed the essays to assess the relationship between these grades and various grammatical attributes. Unfortunately, when we examine ratings from several raters, it generally happens that the classifications assigned to individuals by different raters are not consistent. We must therefore decide how to combine the information gathered from different raters.

Numerous approaches have been proposed for analyzing ordinal data collected from multiple raters. Often, emphasis in such analyses focuses on modeling the agreement between raters. Among the more commonly used indices of multirater agreement in social sciences and medicine is the κ statistic (Cohen 1960). Assuming that all judges employ the same number of rating categories, the κ statistic can be estimated by constructing a contingency table in which each judge is treated as a factor having K levels. The κ statistic is then defined by

$$\kappa = \frac{p_0 - p_c}{1 - p_c},$$

where p_0 represents the sum of the observed proportions in the diagonal elements of the table, and p_c represents the sum of the expected proportions under the null hypothesis of independence. Large positive values of κ may be interpreted as indicating systematic agreement between raters. This statistic has been developed and extended by a number of authors, including Fleiss (1971), Light (1971), and Landis and Koch (1977a,b). A related index has been proposed by Jolayemi (1990a,b).

A more recent, model-based approach toward measuring rater agreement was proposed by Tanner and Young (1985). In this paradigm, the contingency table used in the construction of the κ statistic was analyzed in the context of a log-linear model. Indicator variables corresponding to subsets of diagonal cells in subtables were used to model the agreement between different judges. An advantage of this approach over the κ statistic is that specific patterns of rater agreement can be investigated. Both methodologies are applicable to nominal and ordinal categorical data. Further work in this direction has recently been proposed by Uebersax (1992, 1993) and Uebersax and Grove (1993)

In contrast to these approaches, the approach that we advocate emphasizes the tasks of evaluating rater precision, estimating the relative rankings of individuals, and predicting rankings from observed covariates. Unlike the approaches mentioned above, we assume a priori that all judges essentially agree on the merit of various individuals and that an underlying trait (or trait vector) determines the "true" ranking of an individual in relation to all others. Generally, we assume that this trait is scalar-valued, although a brief discussion of multivariate trait vectors is provided in Chapter 7.

5.1   Essay scores from five raters

To illustrate our modeling approach, let us again consider the essay grade data which we encountered at the end of Chapter 4.

Figure 5.1 depicts the marginal distribution of the grades assigned by each of the five judges to the 198 essays. From this figure, we see that the proportion of essays assigned to each grade category varies substantially from judge to judge. The raters vary with respect to their average ratings and also with respect to the spread of their ratings. For example, Rater 1 appears to assign higher ratings than Rater 2. Rater 3 seems unusual with respect to the relatively large variation of his or her ratings. Of course, the variation between ratings that we see in Figure 5.1 does not necessarily mean that the *rankings* of the essays were not consistent across judges; it might mean only that the grade cutoffs employed by the judges were different. To examine the consistency of the rankings, we can plot the essay grades assigned by the judges against one another. Such a plot is provided in Figure 5.2. In order to make this plot easier to interpret visually, the elements in the cross-

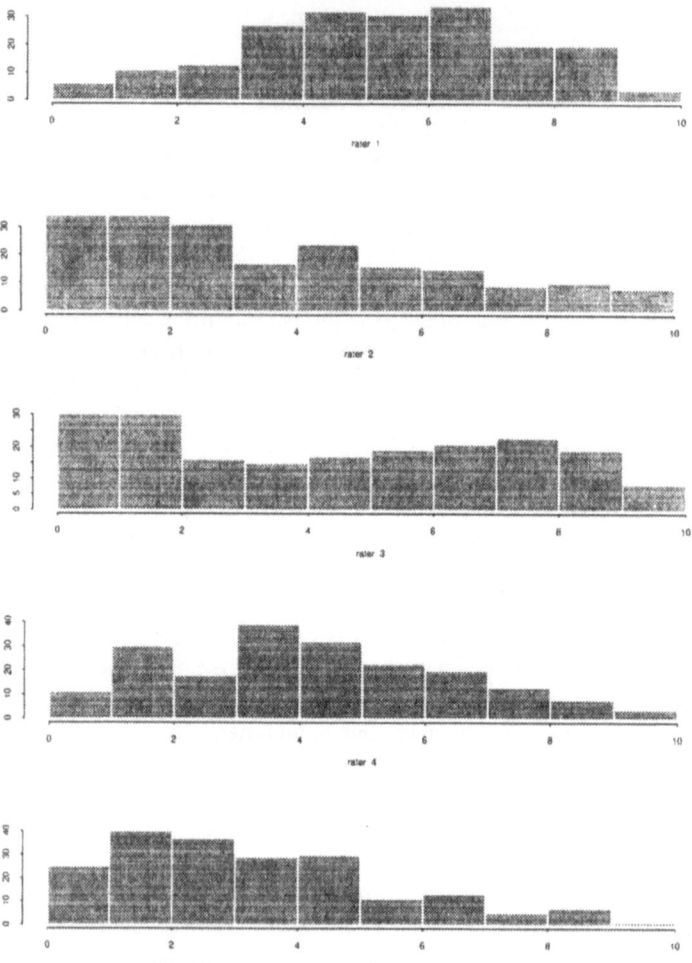

FIGURE 5.1. Histogram estimate of the marginal distribution of the grades assigned by each expert rater to the human essays.

tabulation tables have been plotted as a gray-scale image, with darker squares corresponding to higher counts in the bivariate histogram. The extent to which raters agree is indicated by the concentration of dark squares along a line with positive slope. When raters agree in their rankings of individuals and also employ similar definitions of the category cutoffs, the slope of this line is approximately 1.

From Figure 5.2, we see that the variability of the third rater is comparatively large in comparison to the other four raters. It also appears that the second, fourth, and fifth raters produced rankings that were largely consistent with one another, and that the second and fifth raters used similar category definitions.

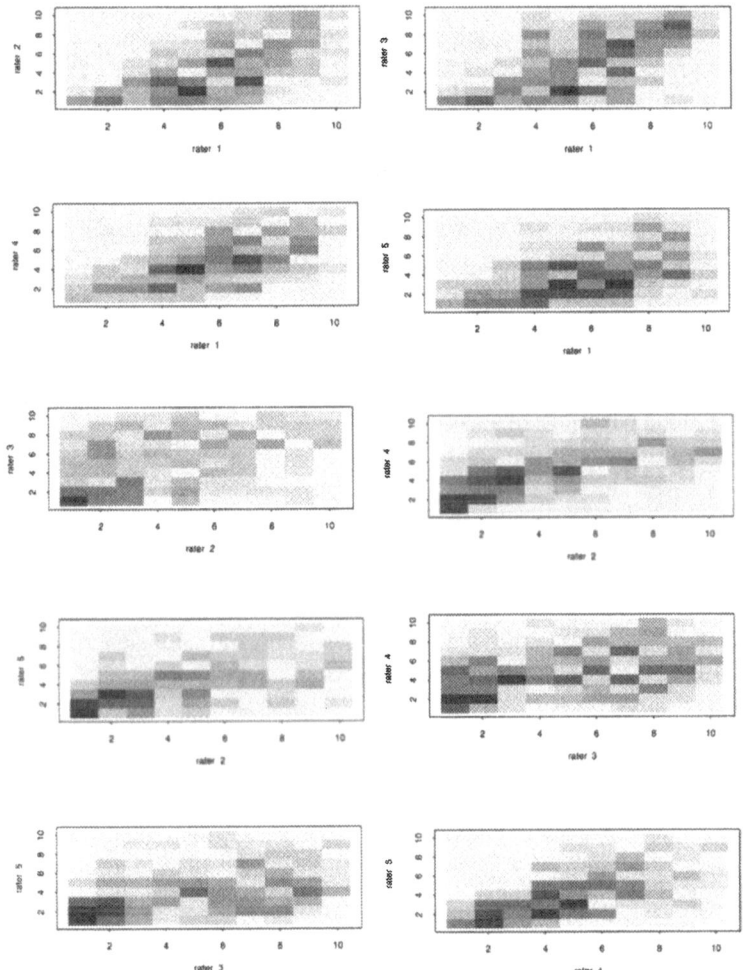

FIGURE 5.2. Bivariate histogram representations of joint marginal distributions of the grades assigned by each pair of expert raters to the human essays. Darker squares represent higher counts.

5.2 The multiple rater model

5.2.1 The likelihood function

As in the preceding chapter, we denote the "true" value of the ith individual's latent trait on a suitably chosen scale by Z_i. The vector of latent traits for all individuals is denoted by $\mathbf{Z} = \{Z_i\}$.

We assume that the available data have the following general form. There are n individuals rated, and each individual is rated by at most J judges or raters. In many cases, every judge rates every individual. In the situation where all individuals are not rated by all judges, we assume that the decision for a judge to rate an individual is made independently of the qualities of both the judge and individual. We further assume that judge j classifies each individual into one of K_j ordered categories. Typically, all judges utilize the same number of categories; in which case, we drop the subscript j and let K denote the common number of ordered categories. We let $\mathbf{y} = \{y_{ij}\}$ denote the data array, with y_{ij} denoting the rating assigned by judge j to individual i. The matrix of covariates relevant for predicting the relative rankings of the individuals is denoted by $\mathbf{X}$.

In assigning a category or grade to the ith object, we assume that judge j observes the value of Z_i with an error denoted by ϵ_{ij}. The quantity $t_{ij} = Z_i + \epsilon_{ij}$ then denotes judge j's perception of the latent trait for individual i on the underlying trait scale. We refer to the quantities $\{t_{ij}\}$ as *perceived traits* or *perceived latent traits*.

The error term ϵ_{ij} incorporates both the observational error of judge j in assessing individual i and the bias of judge j in assessing the true value of Z_i. In some cases, it might be sensible to model ϵ_{ij} as a function of individual covariates, although in what follows we assume that the expectation of ϵ_{ij}, averaged over all individuals in the population that have the same covariate values as individual i is zero.

As in the single rater setting, we assume that individual i is assigned to category c by judge j if

$$\gamma_{j,c-1} < t_{ij} \leq \gamma_{j,c}, \tag{5.1}$$

for judge-specific category cutoffs $\gamma_{j,c-1}$ and $\gamma_{j,c}$. As in Chapter 4, we define $\gamma_{j,0} = -\infty$ and $\gamma_{j,K} = \infty$ and let $\gamma_j = (\gamma_{j,1}, ..., \gamma_{j,K-1})$ denote the vector of cutoffs for the jth judge. Let $\gamma = \{\gamma_1, ... \gamma_J\}$ denote the array of category cutoffs for all judges.

To this point, the model for the multirater ordinal data generation is entirely analogous to the single-rater case. However, in specifying the distribution of the error terms ϵ_{ij}, we must decide whether we wish to assume that all judges rank individuals with equal precision or whether some judges provide rankings that are more accurate than others.

In either case, it is convenient to assume a common distributional form for the error terms ϵ_{ij} across judges. We therefore assume that ϵ_{ij}, the error of the jth judge in perceiving the latent trait of individual i, has a distribution with mean 0 and variance σ_j^2. We write the distribution function of ϵ_{ij} as $F(\epsilon_{ij}/\sigma_j)$ for a known distribution function F. We denote the density function corresponding to F by f.

By taking $\sigma_j^2 = \sigma^2$ for all j, we impose the constraint that all judges rank individuals with similar precision. In practice, however, this assumption is seldom supported by data; therefore, unless explicitly stated otherwise we assume distinct scale parameters for each judge.

Under these assumptions, it follows that the likelihood function for the observed data $\mathbf{y}$ (ignoring, for the moment, regression of the latent traits $\mathbf{Z}$ on explanatory variables **X**) may be written

$$L(\mathbf{Z}, \gamma, \{\sigma_j^2\}) = \prod_{i=1}^{n} \prod_{j \in C_i} \left[F\left(\frac{\gamma_{j,y_{ij}} - Z_i}{\sigma_j} \right) - F\left(\frac{\gamma_{j,y_{ij}-1} - Z_i}{\sigma_j} \right) \right], \quad (5.2)$$

where C_i denotes the set of raters who classified individual i. If we introduce the perceived latent trait values t_{ij} into the likelihood function, the augmented likelihood function can be expressed as

$$L(\mathbf{Z}, \{t_{ij}\}, \gamma, \{\sigma_j^2\}) = \prod_{i=1}^{n} \prod_{j \in C_i} \frac{1}{\sigma_j} f\left(\frac{t_{ij} - Z_i}{\sigma_j} \right) I(\gamma_{j,y_{ij}-1} < t_{ij} \leq \gamma_{j,y_{ij}}). \quad (5.3)$$

As before, $I(\cdot)$ denotes the indicator function. Graphically, this model for the likelihood function is illustrated in Figure 5.3. In this plot, two raters classify an individual with a true latent trait of 1.5, indicated by the isolated vertical line. The distribution of their observations of this individual's trait are depicted by the two normal densities, from which it is clear that the second rater is less precise. The horizontally shaded region represents the probability that the first rater classifies this individual as "2," supposing that the lower and upper category cutoffs for the first rater's second category were $(\gamma_{1.1}, \gamma_{1.2}) = (-1.0, 0.1)$. Similarly, the vertically shaded region depicts the probability that the second rater classified this individual in the second category, given that the corresponding category cutoffs were $(\gamma_{2.1}, \gamma_{2.2}) = (-0.2, 1.0)$.

5.2.2   The prior

Upon careful examination of the likelihood function ((5.2) or (5.3)), it is clear that the model parameters are not identifiable; that is, for any constants a and $b > 0$,

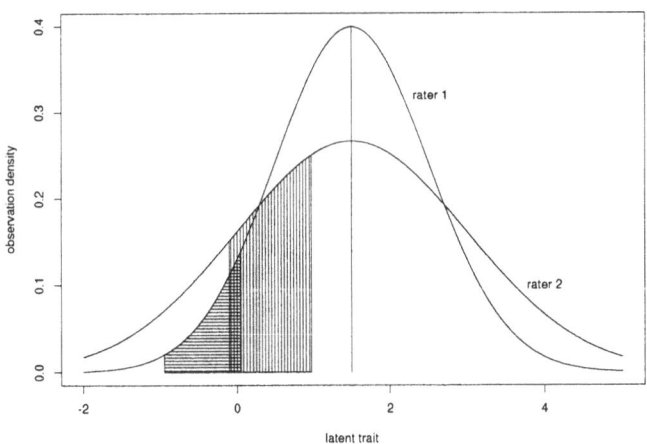

FIGURE 5.3. Depiction of multirater ordinal data model.

we may replace $\mathbf{Z}$ with $b(\mathbf{Z} - a)$, t_{ij} with $b(t_{ij} - a)$, γ with $b(\gamma - a)$, and σ_j with $b\sigma_j$, without changing the value of the likelihood. We faced a less severe identifiability problem in the last chapter when we composed a model for single-rater ordinal data. To solve that problem, we imposed a constraint on the value of the first category cutoff to make γ and the regression intercept identifiable. In this case, the problem is exacerbated because it is generally unreasonable to assume that the upper cutoff for the lowest category is the same for all raters. Furthermore, when data from only one rater are available, the value of the rater variance σ^2 can be assigned the fixed value of 1. This constraint on σ eliminates the scaling problem (i.e., multiplying all model parameters by a positive constant b), but as stated above, it is generally not appropriate for multirater data due to the fact that different raters exhibit different levels of precision in their rankings.

The identifiability problem can be overcome by imposing proper prior distributions on some or all model parameters. As in Chapter 4, the location of the trait distribution can be fixed by specifying a proper prior for the latent traits. For convenience, we assume throughout this chapter that the proper prior chosen for the latent traits is a Gaussian distribution; in other words, that the latent traits $Z_1, ..., Z_n$, are distributed a priori from independent standard normal distributions.

In addition to specifying a proper prior distribution on the latent trait vector $\mathbf{Z}$, we also assume a specific distributional form for the rater variance parameters σ_j^2. Recalling that rescaled versions of the rater error terms were assumed to be distributed according to a known distribution F, we assign a proper prior distribution to the rater variances, which determine the scaling factors of the rater errors. In other words, we assume that the rater error terms ϵ_{ij}/σ_j are distributed according to F, and take an informative prior on σ_j^2. A convenient choice for $F(\cdot)$ is a standard normal distribution. If we combine this assumption with the assumptions made above, it follows that the conditional distribution of the perceived latent traits $\{t_{ij}\}$, given Z_i, are independent and normally distributed with with mean Z_i and variance σ_j^2.

The assumption of normality of the judge error terms can be at least partially justified by noting that the errors in a judge's perception of an individual's attributes usually result from a large number of small effects. By the central limit theorem, we might therefore expect that the rater errors are approximately Gaussian. Furthermore, it should be noted that predictions obtained under a model that assumes normally distributed rater errors generally produces predictions that are quite similar to predictions obtained under other common error models. Thus, the final conclusions drawn from this class of models tend to be relatively insensitive to the particular distributional form assumed for the components of ϵ_{ij}.

A priori, we also assume that the rater variances $\sigma_1^2, ..., \sigma_J^2$ are independent. If F is chosen to be a standard normal distribution, then the conjugate prior for a variance parameter σ_j^2 is an inverse-gamma density, expressible in the form

$$g(\sigma_j^2; \lambda, \alpha) = \frac{\lambda^\alpha}{\Gamma(\alpha)} (\sigma_j^2)^{-\alpha-1} \exp\left(-\frac{\lambda}{\sigma_j^2}\right), \qquad \alpha, \lambda > 0. \qquad (5.4)$$

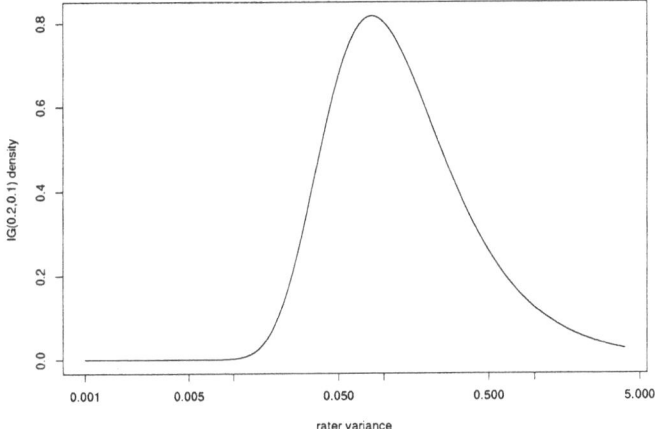

FIGURE 5.4. Inverse-gamma density with parameters $\alpha = 0.2$ and $\lambda = 0.1$. Note that the x axis is plotted on the logarithmic scale to better illustrate the behavior of the density near 0.

We denote the inverse-gamma distribution corresponding to this density by $IG(\alpha, \lambda)$; the mean and mode of the distribution are $\lambda/(\alpha - 1)$ (assuming $\alpha > 1$) and $\lambda/(\alpha + 1)$, respectively. A density plot for an $IG(0.2, 0.1)$ random variable is depicted in Figure 5.4. The parameters $\alpha = 0.2$ and $\lambda = 0.1$ were chosen so that the prior density on the rater variances concentrates its mass in the interval $(0.01, 4.0)$. It is important to assign a positive value to λ in order to prevent singularities in the posterior distribution that would occur if the components of σ^2 were allowed to become arbitrarily small.

To complete the prior model, we need to specify a distribution on the vector of category cutoffs γ. For present purposes, we assign independent uniform priors on the category cutoff vectors γ_j, subject to the constraint that

$$\gamma_{j,1} \leq \cdots \leq \gamma_{j,K-1}.$$

Combining all of these assumptions, the joint prior density on $(\mathbf{Z}, \gamma, \{\sigma_j^2\})$ is given by

$$g(\mathbf{Z}, \gamma, \{\sigma_j^2\}) = \prod_{i=1}^{n} \phi(Z_i; 0, 1) \prod_{j=1}^{J} g(\sigma_j^2; \lambda, \alpha), \tag{5.5}$$

where $\phi(x; \mu, \sigma^2)$ denotes a normal density with mean μ and variance σ^2. Taken together, this set of assumptions defines what we refer to as the multirater ordinal probit model. As we will see in later chapters, this model provides a useful framework for analyzing a wide variety of ordinal datasets.

5.2.3   Analysis of essay scores from five raters (without regression)

To quantify the qualitative conclusions drawn from Figures 5.1 and 5.2, we apply the model described in the last section to obtain the posterior distributions on each rater's variance parameter. As a by-product of this model fitting procedure, we also obtain the posterior distribution on the underlying trait for each essay's grade.

With the introduction of the latent trait estimates t_{ij} into the estimation problem, the joint posterior density of all unknown parameters is given by

$$g(\mathbf{Z}, \{t_{ij}\}, \gamma, \{\sigma_j^2\}|\mathbf{y}) \propto L(\mathbf{Z}, \{t_{ij}\}, \gamma, \{\sigma_j^2\})g(\mathbf{Z}, \gamma, \{\sigma_j^2\}),$$

where the likelihood function is given by (5.3) and the prior density by (5.5). To obtain samples from this posterior distribution, we modify the MCMC algorithm described in the previous chapter for single-rater data to accommodate additional raters. After initializing model parameters, we begin the MCMC algorithm by sampling from the conditional distribution of $\mathbf{Z}$. From (5.3), we see that the conditional distributional of the component Z_i, given the array $\{t_{ij}\}$ and σ_j^2, is normally distributed with mean s/r and variance $1/r$, where

$$r = 1 + \sum_{j \in C_i} \frac{1}{\sigma_j^2} \quad \text{and} \quad s = \sum_{j \in C_i} \frac{t_{ij}}{\sigma_j^2}. \tag{5.6}$$

Given the value of $\mathbf{Z}$, updating the components of γ_j for each rater j may be accomplished in precisely the same manner that updates to a single rater's category cutoffs were performed in Section 4.3.2. Similarly, perceived trait values t_{ij} can be sampled from a truncated Gaussian density with mean Z_i and variance σ_j^2, truncated to the interval $(\gamma_{j,y_{ij}-1}, \gamma_{j,y_{ij}})$.

Finally, the conjugate prior structure specified for the variances $\sigma_1^2, ..., \sigma_J^2$ makes sampling from the conditional distributions of these parameters straightforward. If we let D_j denote the set of individuals rated by the jth judge and take n_j to be the number of elements of D_j, then the conditional distribution of σ_j^2 is

$$\sigma_j^2 \sim \mathrm{IG}\left(\frac{n_j}{2} + \alpha, \frac{S}{2} + \lambda\right) \quad \text{where} \quad S = \sum_{i \in D_j}(t_{ij} - Z_i)^2. \tag{5.7}$$

These conditional distributions can be used to sample from the joint posterior distribution on all model parameters. After this MCMC algorithm is implemented, we can estimate the posterior means and standard devations of the rater variance parameters from the MCMC output. Estimates obtained in this way are depicted in Table 5.1. Note that the values displayed in this table agree qualitatively with the graphical analysis of the data presented in Figure 5.2. As predicted, the third rater tended to assign essay grades that were not consistent with the grades assigned by the other raters.

	Rater				
	1	2	3	4	5
Posterior mean	0.91	0.53	2.05	0.61	0.89
Standard deviation	0.22	0.14	0.54	0.14	0.24

TABLE 5.1. Posterior means and posterior standard deviations of rater variance parameters.

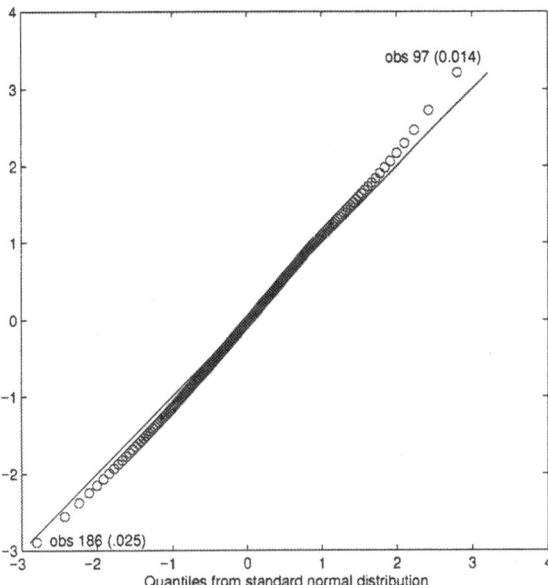

FIGURE 5.5. Normal scores plot for the posterior means of the standardized ordered latent residuals for the first rater's grades.

The values obtained from the MCMC algorithm can also be used to perform residual analyses in ways similar to those described for single-rater data in Chapter 4. For example, the simulated values of t_{ij} and $\mathbf{Z}$ obtained from the MCMC algorithm can be used to define standardized residuals of the form

$$r_{ij} = \frac{t_{ij} - Z_i}{\sigma_j}.$$

A normal scores plot of the posterior means of the ordered standardized residuals for the first rater's grades is shown in Figure 5.5. In this case, the ordered standardized results fall approximately on a line with slope 1 and intercept 0, suggesting that none of the grades assigned by this rater were abnormally deviant.

5.3 Incorporating regression functions into multirater data

If we compare the model framework outlined above for multirater ordinal data to the standard model for single-rater ordinal data described in Chapter 4, we find two basic differences. First, in the case of a single-rater data, there is no loss of generality incurred by fixing the value of the first category cutoff at zero, provided that we include an intercept in the linear regression of the latent trait variables $\{Z_i\}$ on the matrix of explanatory variables $\mathbf{X}$. Second, and more important, is the fact that we implicitly assumed a value of 0 for the rater-variance parameter in the single-rater case. Coupled with the assumption that

$$Z_i = \mathbf{x}'_i \beta + \eta_i, \qquad\qquad \eta_i \sim F(\cdot), \qquad\qquad (5.8)$$

where $F(\cdot)$ denoted the link function for the regression model, this allowed us to define a scale of measurement for both the latent variables and the regression parameter. Of course, with data from only one rater, we really had no choice but to make the assumption that the rater correctly categorized each observation; that is, that the rater's error variance was exactly 0. Indeed, in many instances this assumption might actually be justified from substantive considerations. In testing mechanical parts for failure, for example, the binary classification of tested parts as either a success or failure might be completely objective.

However, for multirater data, the situation changes. The very fact that data in an experiment or study were collected from multiple raters implies that the classification of individuals into categories was subjective; that is, different raters are *expected* to have different opinions on the relative merit of each individual. Substantively, the subjectiveness of the observed data means that one must question the validity of the regression assumption.

To illustrate the importance of this point, recall that in the case of single-rater data, the latent traits Z_i were assumed to follow the regression relation (5.8). If we assume that $F(\cdot)$ is a standard normal distribution function, as we do throughout this chapter, Equation (5.8) can be combined with the model assumptions of the previous section to obtain the following expression for the perceived latent trait observed by a single rater:

$$t_{ij} = \mathbf{x}'_i \beta + \eta_i + \epsilon_{ij} \quad \text{where} \quad \eta_i \sim N(0, 1), \quad \text{and} \quad \epsilon_{ij} \sim N(0, \sigma_j^2). \quad (5.9)$$

It follows that the perceived latent trait t_{ij} has a normal distribution with mean $\mathbf{x}'_i \beta$ and variance $1 + \sigma_j^2$. This implies that the estimated variance for those raters whose classifications most closely follow the regression function will be the smallest. Data obtained from these raters will consequently be given more weight than the other raters in estimating the true ranking of individuals.

These ideas are well illustrated in terms of our example involving the the essay grades collected independently from five judges. In the previous chapter, we posited a linear model for the latent performance variable associated with the grade acquired from the first judge. Assuming that a similar model can be used to predict the grade of any of the five judges participating in the study, we obtain the following regression equation for the prediction of t_{ij}:

$$t_{ij} = \beta_0 + \beta_{WL}\text{WL}_i + \beta_{SqW}\text{SqW}_i + \beta_{PP}\text{PP}_i + \eta_i + \epsilon_{ij}. \quad (5.10)$$

As before, WL, SqW, and PP represent the average word length, the square root of the number of words in the essay, and the percentage of prepositions used, respectively. The variance of ϵ_{ij}, σ_j^2, measures the agreement of the jth judge's ratings with the explanatory variables. From a substantive viewpoint, the critical question is, "Do we wish to give more weight to the rankings of those judges whose grades were most linear in these explanatory variables?" In this example, the answer to this question is, "probably not." Our primary interest in performing this regression was to investigate the extent to which an expert's grade might be modeled using easily quantified grammatical variables. We did not anticipate that these variables would ideally predict the relative merit of the essays.

The above discussion motivates the consideration of a more general model that can accomodate the lack of fit of the regression model. One way to do this is to assume that for given values of the parameters β and τ^2,

$$Z_i = \mathbf{x}_i'\beta + \eta_i + \zeta_i, \qquad \text{where} \quad \eta_i \sim N(0, 1), \quad \text{and} \quad \zeta_i \sim N(0, \tau^2); \quad (5.11)$$

that is, we put the regression equation on equal footing with the ratings obtained from a single judge. The error term η_i provides the model link function, or the randomness associated with categorizing an individual into a given class, and ζ_i accounts for the "lack-of-fit" error associated with the regression equation. This term is completely analogous to the term ϵ_{ij} associated with the perceived latent traits $\{t_{ij}\}$. The precision of the regression relationship depends on the value of τ^2.

The difficulty with formulation (5.11) is that it is inconsistent with the assumptions made in the last section concerning the marginal distribution on $\mathbf{Z}$—that $\mathbf{Z} \sim N(0, \mathbf{I})$. Recall that this assumption was needed to make parameters in the likelihood identifiable. A similar identifiability problem also exists here. Unfortunately, if a vague prior is specified for β, constraint (5.11) does not establish a scale of measurement for the latent traits.

To summarize our discussion to this point, we have reached two conclusions. First, in many applications, it is not reasonable to assume that the "true" rating of an individual is exactly predicted by the regression function. For this reason, the assumptions implied by model (5.9) are often inappropriate for modeling ordinal data. Second, we cannot assume that the values of the latent trait vector $\mathbf{Z}$ are governed by a relationship of the type expressed in (5.11) without assuming a proper prior on the components of β. Doing so leads to an improper prior on $\mathbf{Z}$ and nonidentifiability of all model parameters.

We can solve each of these problems by binding our ordinal regression model to the multirater ordinal model described in Section 5.2. Essentially, this means that we must specify the conditional distribution of β given $\mathbf{Z}$, rather than the reversing the conditionality relationships and specifying the distribution of $\mathbf{Z}$ given β. In this way, we can preserve the prior assumption that $\mathbf{Z} \sim N(0, \mathbf{I})$ while coupling the scale of β to the scale used to model the rater variances.

To specify the conditional distribution of β given $\mathbf{Z}$, we use standard results from the Bayesian analysis of the normal linear model. In the normal setting, if we let $\mathbf{W}$ denote the data vector and assume that $\mathbf{W} \sim N(\mathbf{Xb}, a^2\mathbf{I})$ for an unknown regression parameter $\mathbf{b}$ and known variance a^2, then the posterior distribution of $\mathbf{b}$ is

$$\mathbf{b} \sim N\left((\mathbf{X}'\mathbf{X})^{-1}\mathbf{X}'\mathbf{W}, a^2(\mathbf{X}'\mathbf{X})^{-1}\right),$$

when a uniform prior is assumed for $\mathbf{b}$. If the prior on $\mathbf{b}$, given a^2, is $N(d, a^2 D)$, then the posterior distribution for $\mathbf{b}$ is

$$\mathbf{b} \sim N(f, a^2 F),$$

where

$$f = \left[(\mathbf{X'X}) + D^{-1} \right]^{-1} (\mathbf{X'W} + D^{-1}d)$$

and

$$F = (\mathbf{X'X} + D^{-1})^{-1}.$$

Applying these normal theory results to our problem, we might therefore assume that the *prior* distribution for β, *given* $\mathbf{Z}$ and τ^2, can be expressed as

$$\beta \mid \mathbf{Z}, \mathbf{X}, \tau^2 \sim N(c, C). \tag{5.12}$$

In the absence of specific prior information regarding the prior density for β, we take c and C to be the least squares estimates:

$$c = (\mathbf{X'X})^{-1}\mathbf{X'Z} \tag{5.13}$$

and

$$C = \tau^2(\mathbf{X'X})^{-1}. \tag{5.14}$$

Alternatively, when prior information concerning the regression parameter β is available, the parameters c and C might be chosen as

$$c = \left[(\mathbf{X'X}) + D^{-1} \right]^{-1} (\mathbf{X'Z} + D^{-1}d),$$

$$C = \tau^2(\mathbf{X'X} + D^{-1})^{-1},$$

where d and $\tau^2 D$ are the prior mean and covariance of β, respectively.

Continuing the analogy with the normal-theory models, we complete our specification of the prior model by taking the prior for τ^2, given $\mathbf{Z}$, to be an inverse-gamma density of the form

$$g(\tau^2 \mid \mathbf{Z}) = \frac{(\mathrm{RSS}/2 + \lambda_r)^{(n-p)/2+\alpha_r}}{\Gamma[(n-p)/2 + \alpha_r]} \ (\tau^2)^{-(n-p)/2-\alpha_r-1}$$

$$\times \exp\left(-\frac{\mathrm{RSS}/2 + \lambda_r}{\tau^2} \right). \tag{5.15}$$

In (5.15), RSS denotes the residual sum of squares of the regression of $\mathbf{Z}$ on $\mathbf{X}\beta$; that is, $\mathrm{RSS} = \mathbf{Z'}(\mathbf{I} - \mathbf{X'}(\mathbf{X'X})^{-1}\mathbf{X})\mathbf{Z}$, and α_r and λ_r are prior hyperparameters.

From a technical standpoint, it is interesting to note that this specification of the prior density for β and τ^2 (given $\mathbf{Z}$) differs from the prior density implied by model (5.11) by a factor of

$$(\mathrm{RSS}/2 + \lambda_r)^{(n-p)/2+\alpha_r}.$$

In the case of a vague prior for β and τ^2 (that is, when $\lambda_r = \alpha_r = 0$), this factor approaches zero as the residual sum of squares approaches zero. Multiplication by this factor in the revised model prevents the posterior distribution from becoming arbitrarily large in a region near the value $\tau^2 = 0$, since the residual sum of squares approaches 0 as $\mathbf{Z}$ becomes small. Without this factor, the prior specified in (5.11) leads to a posterior distribution that is unbounded for small values of $\mathbf{Z}$ and τ^2.

Finally, we must specify values for the hyperparameters α_r and λ_r. A natural choice for these parameters is to assign them the values α and λ used in the specification of the prior on the rater variances. Doing so facilitates the comparison of the posterior distribution on the rater variances and the variance of the regression relation. Alternatively, the prior on τ^2 might be taken to be a scaled version of the prior assumed for σ_j^2, with a scaling constant determined from prior knowledge of the relative reliability of a single judge's scores relative to predictions obtained from the regression relation.

5.3.1 Regression of essay grades obtained from five raters

We now reexamine the regression of the essay grades from the five raters described earlier. At the end of Chapter 4, we explored the relationship between the grades assigned to the essays by the first judge and the grammatical characteristics of the average word length (WL), the square root of the number of words in the essay (SqW), the average sentence length (SL), and the percentages of misspellings (PS), commas (PC), and prepositions (PP). Of these six variables, only SqW, WL, and PP appeared to be important in predicting the first rater's grades.

To explore possible relationships between the grammatical characteristics of the essays and the five grades obtained for each essay, it is convenient to use as a summary measure of an essay's quality the posterior mean of the latent trait vector $\mathbf{Z}$ estimated in Section 5.2.3. Scatterplots of the estimated posterior means of $\mathbf{Z}$ against each of the grammatical variables are depicted in Figure 5.6.

The posterior means of the latent traits displayed in Figure 5.6 may also be used to perform informal regression analyses prior to generating posterior samples from the final model. Practically speaking, performing such analyses can result in substantial time savings, as the need to sample from the posterior distribution of a large number of potential regression models is eliminated.

To illustrate this strategy, the estimated posterior mean of the latent trait vector $\mathbf{Z}$ was regressed on the six grammar variables depicted in Figure 5.6. Weighted regression was performed, using a diagonal weight matrix with weights inversely proportional to the posterior variances of the latent trait vector. For computational simplicity, we ignored the posterior correlations of the components of $\mathbf{Z}$. Results from the regression on all six essay characteristics are provided in Table 5.2.

With the exception of average sentence length (SL), all variables in the regression now appear to be significant. Recall that in a similar regression on data from the first rater, only the variables SqW, WL, and PP appeared significant, and PP was only marginally so.

An informal significance test of the adequacy of the regression in predicting the latent trait vector $\mathbf{Z}$ can be performed by testing whether the estimate of the mean squared error s^2 obtained from the weighted least-squares regression could plausibly be equal to 1. If so, then this suggests that the regression function adequately predicted the values of $\mathbf{Z}$, within the range of posterior uncertainty reflected in this

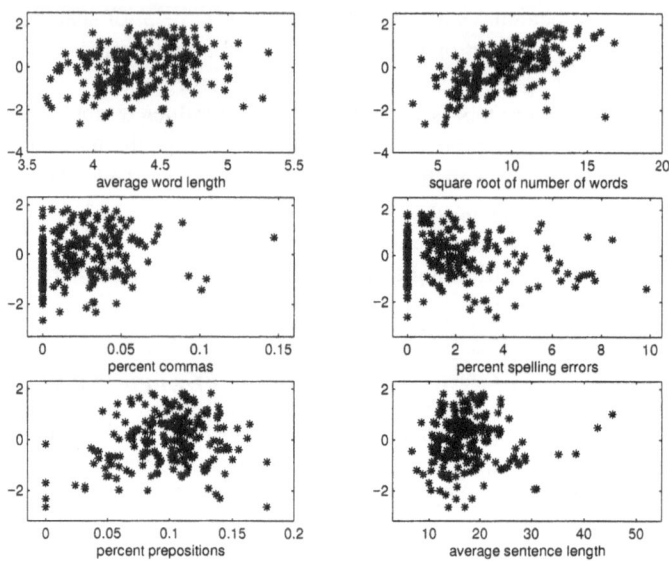

FIGURE 5.6. Scatterplots of estimated essay latent traits versus six essay characteristics.

	Explanatory Variable						
	Intercept	WL	SqW	PC	PS	PP	SL
LSE	−5.34	0.70	0.21	5.29	−.13	3.99	0.0005
Std. dev.	0.72	0.16	0.020	2.19	0.026	1.59	.0088
t-value	−7.45	4.51	10.72	2.41	−4.83	2.50	0.06

TABLE 5.2. Weighted least-squares estimates (LSE) from regression of the posterior mean of $\mathbf{Z}$ on six grammatical variables. The weight matrix was assumed to be a diagonal matrix with diagonal entries equal to the inverse of the marginal posterior variance of components of $\mathbf{Z}$.

parameter vector. If not, there is evidence of model lack of fit. In this case, the residual standard error of the weighted least-squares regression was 1.782 on 191 degrees of freedom, so that the mean square error estimate is $\hat{s}^2 = 3.17$. Under the null hypothesis that the observational variance is given by the posterior uncertainty in $\mathbf{Z}$, $\hat{s}^2$ is distributed as a $\chi^2_{191}/191$ random variable. It follows that the probability of observing a value of $\hat{s}^2$ greater than 3.17 under the null hypothesis is less than .0001. Thus, as we suspected a priori, the grammatical variables are clearly not adequate in predicting the underlying merit of an essay.

Based on the regression results displayed in Table 5.2, we assume a final regression model for the essay grade data of the form

$$\left(\beta_0, \beta_{WL}, \beta_{SqW}, \beta_{PS}, \beta_{PC}, \beta_{PP}\right)' \mid \mathbf{Z}, \tau^2 \sim N\left((\mathbf{X}'\mathbf{X})^{-1}\mathbf{X}'\mathbf{Z}, \tau^2(\mathbf{X}'\mathbf{X})^{-1}\right),$$
(5.16)

where $\mathbf{X}$ denotes the design matrix which has as its columns a vector of 1's and the variables *WL, SqW, PS, PC* and *PP*. The likelihood function for the essay grades is as specified in (5.2), or equivalently (5.3). To facilitate comparisons between the rater variances $\{\sigma_j^2\}$ and the variance parameter τ^2 associated with the regression function, we assume a priori that all variance parameters are distributed independently according to an inverse-gamma density with parameters $\alpha = 0.2$ and $\lambda = 0.1$. To complete the model specification, we also assume that the components of $\mathbf{Z}$ are distributed independently a priori as standard normal deviates.

Computational strategies

The MCMC algorithm described in Section 5.2.3 can be modified in a straightforward way to accommodate the addition of the regression function. Because the conditional distributions of the rater-specific category cutoffs, given the rater variances and the value of $\mathbf{Z}$, are not affected by the inclusion of the regression function, MCMC updates of the category cutoffs can be performed as before. Similarly, given $\mathbf{Z}$ and the rater variances, the conditional distributions of the perceived latent traits $\{t_{ij}\}$ are unchanged by the introduction of the regression function. Likewise the conditional distributions of the rater variances given the components of t_{ij} and $\mathbf{Z}$ are the same in the nonregression case. The only changes to the basic MCMC algorithm used to sample from the posterior distribution on the model parameters involve the updating strategies used to sample from the components of $\mathbf{Z}$ and τ^2.

To sample from the conditional distributions of the components of $\mathbf{Z}$ given all other model parameters, the factors multiplying the posterior that arise from the inverse-gamma density on τ^2 and the normal distribution on β must be included. Although it is possible to introduce an additional set of latent variables for the regression equation that are similar to the perceived latent traits $\{t_{ij}\}$, it is perhaps easier to update $\mathbf{Z}$ using a Metropolis-Hastings update. This update can be implemented using the conditional distribution of $\mathbf{Z}$ given in (5.6) as a proposal density. With this proposal, the ratio of the products of the inverse-gamma density on τ^2 and the multivariate normal density specified on β ((5.13)–(5.14)), evaluated at the candidate and current values of $\mathbf{Z}$, provides the appropriate accept/reject probability. In general, this acceptance probability is typically close to 1, and so little efficiency is lost if this sampling strategy is adopted.

Sampling from the conditional distributions of β and τ^2 is straightforward using Gibbs updates based on the full and reduced conditionals used in the model specification.

Numerical Results

Table 5.3 displays the posterior means for the regression parameters of model (5.3). Least-squares estimates obtained by fitting weighted least squares to the posterior means of the latent traits, as described above but without the variable SL, are also provided.

As this table indicates, the values of the posterior mean of the parameter estimates in the two models are quite similar, although the standard errors of the least-squares estimates tend to be smaller. Ignoring differences in interpretation

		Explanatory variable				
	Intercept	WL	SqW	PC	PS	PP
Posterior mean	−5.46	0.72	0.22	5.99	−0.13	3.90
Std. dev.	0.94	0.21	0.028	2.97	0.035	2.03
Least-squares	−5.33	0.70	0.21	5.30	−0.13	4.01
Std. dev.	0.70	0.16	0.020	2.19	0.026	1.56

TABLE 5.3. Posterior means and posterior standard deviations of the regression parameters. The least-squares estimates obtained by fitting the posterior mean estimates of the latent traits without the regression model are provided for comparison.

between the classical and Bayesian estimates, much of the underestimation of the classical standard errors results from the fact that the least-squares estimates were obtained through a conditional analysis in which the response variable was held fixed, while uncertainty in the response was fully accounted for in the Bayesian model. In addition, some of the differences in the parameter estimates may be attributed to the fact that the regression function exerted some influence on the estimation of the latent traits Z in the Bayesian model (recall that the conditional distribution used to update Z changed when the regression relation was included in the model).

Table 5.4 illustrates the effect on the rater variance parameters of including the regression model in the statistical model for the latent traits. Despite the introduction of the regression variance parameter, the estimates of the rater variances for raters whose essays rankings fell near the regression surface were still slightly decreased when the regression relation was formally included in model framework. Conversely, the variance parameters of those raters whose essay grades did not correlate as highly with the regression surface were slightly inflated.

As an interesting side note, the regression model appears to outperform three of the five raters in this example, although it contains only five relatively easily quantified grammatical characteristics.

5.4   ROC analysis

Receiver operator characteristic analysis (ROC) is a technique used in medicine, engineering, and psychology to assess the performance of diagnostic systems and signal recovery techniques (e.g., Swets, 1979; McNeil and Hanley, 1984; Metz,

	Rater variance					Regression
	1	2	3	4	5	variance (τ^2)
Post. means (w/reg.)	0.81	0.55	2.41	0.60	0.95	0.70
Std. dev.	0.20	0.16	0.78	0.13	0.23	0.10
Post. means (w/o reg.)	0.91	0.53	2.05	0.61	0.89	—
Std. dev.	0.22	0.14	0.54	0.14	0.24	—

TABLE 5.4. Posterior means of the rater variance parameters under the model including the regression relation (w/reg.) and without. The indicated standard deviations are the estimated posterior standard deviations of the variance parameters.

1986). It is most often employed to assess the performance of two or more procedures in distinguishing two populations of individuals, often labeled "disease" and "disease-free." The analysis is based on one or more raters evaluating individuals from each group on an ordinal scale. For this reason, ROC analysis may be regarded as a slight generalization of the multirater analysis described earlier. The generalization is that individuals are assumed to be drawn from two (normal) populations rather than one, requiring that two additional model parameters—the mean and variance of the second population—be estimated from the data.

An example of a setting in which ROC analysis might be applied is the evaluation of a new imaging system. For instance, suppose that we wished to compare a new X-ray system to an older one. Although the technical attributes of the two systems might suggest that the new system was superior to the old, in practice physicians might be interested in more direct measures of system performance. For example, they might wish to know which of the two systems was most effective in detecting lung nodules in patients suspected of having lung cancer. To compare the two systems using this criterion, it would be necessary to conduct a study in which chest X-rays were collected from patients on both systems. Data from such a study could be used to choose between the two systems, provided, of course, that accurate diagnoses of each patients condition could be ascertained precisely using other criteria.

As in most testing problems, two types of errors must be balanced against one another in assessing the success of a diagnostic system, namely incorrectly classifying diseased individuals as healthy, and classifying healthy individuals as diseased. The probability of the former is usually denoted FNF (false negative fraction), whereas the probability of the latter is denoted FPF (false positive fraction). In ROC analyses, the trade-off between these two types of errors is summarized using an ROC curve.

An ROC curve is formed by plotting the TPF (true positive fraction, TPF = $1 - $ FNF) versus the FPF as an underlying threshold defining disease status is varied. The resulting plot is defined on the unit square, and the area under the curve, usually denoted A_Z, represents one measure of the utility of the diagnostic system. Values close to 1 are desirable, since such values can be obtained only when the TPF is close to 1 even when the FPF is small. In other words, when the ROC area is close to 1, there is a high probability that a diseased individual will be diagnosed as diseased. At the same time, there is a small probability that a disease-free individual will be diagnosed as diseased. A hypothetical ROC curve is illustrated in Figure 5.7.

Numerous experimental designs can be employed to collect data for estimating ROC curves. The most common design, the so-called rating method, requires that each of several judges rate individuals on an ordinal scale, often into one of four or five categories. The labels assigned to each category range from definitely or almost definitely disease-free (lowest) to definitely or almost definitely diseased (highest).

Given this type of rating data, ROC curves can be constructed in several ways. The prevalent method in the biomedical literature is to assume that individual ratings are based on unobserved Gaussian deviates. In this paradigm, the distri-

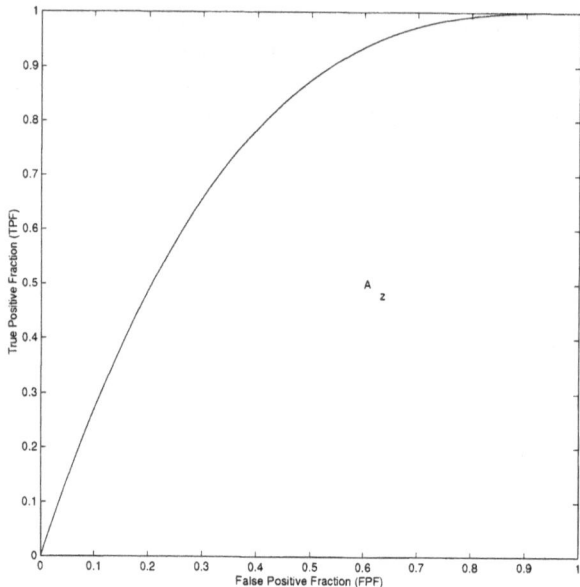

FIGURE 5.7. Hypothetical ROC plot. The area under the curve, A_Z, is often used as a global measure of the value of a diagnostic system.

bution of latent traits for individuals drawn from the disease group is assumed to be $N(a, b)$, whereas disease-free traits are assumed to be drawn from a standard normal distribution, $N(0, 1)$. This set of assumptions is called the binormal ROC model. Figure 5.8 illustrates these assumptions, which we see are very similar to the multirater ordinal probit model.

An important difficulty encountered by classical practitioners of ROC analysis involves combining information obtained from several judges. This difficulty is caused by the absence of a model component for rater variability, and two alternative methods are commonly used to overcome it (see, for example, Metz (1986)). In the first method, ratings from all judges are simply pooled to form a single dataset, as if a single rater had produced all rankings, and the binormal model parameters are estimated as if the data had been generated from this single judge. This is a particularly poor approach toward resolving the problem, since different raters seldom utilize similar category cutoffs. Combining data in this way often leads to a serious loss of information.

In the second approach, model parameters, or derived quantities like A_Z, are estimated separately from each rater's data and are then combined to form pooled estimates of the parameters of interest. This approach tends to provide better estimates of model parameters than the first, but is inefficient due to the fact that information gleaned from comparatively unreliable raters is given the same weight as information obtained from highly expert raters.

The model described in Section 5.2 can be easily extended to accommodate the binormal ROC model. It also provides a natural way of combining information col-

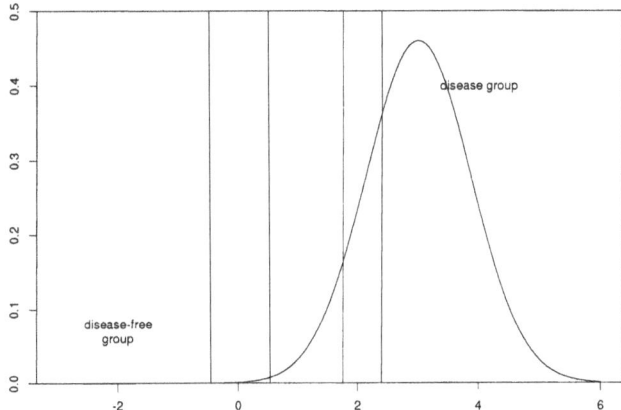

FIGURE 5.8. Graphical depiction of binormal ROC assumptions. Deviates from disease and disease-free groups are assumed to be generated from the normal densities depicted. Vertical bars indicate hypothetical rater thresholds for assignments of disease status to five categories. In this figure, latent traits from the disease group are assumed to be drawn from a normal distribution with mean 3 and variance .64 (i.e., $a = 3, b = .64$)

lected from multiple judges. Conditional upon disease status, the prior distribution on the latent trait for an individual (image) Z_i is assumed to be either $N(0, 1)$ or $N(a, b)$, depending on whether the individual was disease-free or diseased. Since each individual's disease status is assumed known, the only technical adaptation required for model implementation is the conditional estimation of the parameters a and b for the disease group. However, given the disease scores Z_i, estimating (a, b) is tantamount to estimating the mean and variance from a sample of normal observations. Thus, any MCMC algorithm used to implement the hierarchical model of Section 5.2 can be readily adapted to estimate the parameters of interest in ROC analyses.

The latent variable approach described in this chapter offers several benefits over standard ROC analysis. One important advantage of the Bayesian approach is that the effects of rater variability are separated from the efficacy of the diagnostic systems under investigation. Under the Bayesian model, estimated values of the parameters a, b, and A_Z represent the values that would be obtained from an *ideal* observer and are thus tied more directly to the performance of the diagnostic system. In addition, individual ratings and rater performances may be explored as functions of underlying covariates using the techniques described in Section 5.3. Also, the posterior distributions of quantities like A_Z are available directly, and assumptions regarding asymptotic normality of maximum likelihood estimates need not be made. Thus, the posterior probability that one system's A_Z exceeds the other's can be computed exactly. Finally, pooling observations from different observers occurs automatically within the model framework.

As an illustration, data were obtained from an ROC study comparing the ability of observers to detect tumors in single photon emission computed tomography (SPECT) images (Li et al., 1991). SPECT is a medical imaging modality used to image metabolic activities through cross sections of a patient's body, and data for SPECT can be collected using numerous collimator designs. Two collimator designs were compared in this study, parallel beam (PB) and cone beam (CB). Images were obtained for 480 simulated patients, 240 with disease and 240 without. Four judges rated each image on a 5-point scale, and the data were analyzed by Li et al. using the binormal assumptions and the second pooling method described above. The maximum likelihood estimates of the ROC curve area, A_Z, was reported to be 0.90 for CB and 0.82 for PB. Note that these figures represent the averaged areas obtained for the raters selected for this particular study, not an "ideal" rater.

For comparison, we also fit the multirater model described in Section 5.2 to the same data. Normal priors with mean 0 and variance 9 were assumed for the location parameter a of the disease groups' latent trait distribution (under both CB and PB image acquisition treatments), and inverse-gamma densities with parameters $\alpha = \lambda = 5$ were assumed for the variance b of these distributions. The prior on a represents our knowledge that the separation of the two distributions is not so large as to make an ROC study superfluous, and the prior on the variance parameter reflects our knowledge that the variation in reconstructed images obtained from the disease and nondiseased populations should be similar. The prior mean and mode of this inverse-gamma distribution are .87 and 1.25, respectively.

Combining all of model assumptions, it follows that the joint posterior distribution for each treatment's data is proportional to

$$g(\{t_{ij}\}, \mathbf{Z}, \gamma, \sigma_j^2, a, b | \text{data})$$

$$\propto \prod_{j=1}^{J} \prod_{i \in T_j} \left\{ \frac{1}{\sigma_j} \exp\left[-\frac{(t_{ij} - Z_i)^2)}{2\sigma_j^2} \right] I(\gamma_{j, y_{ij}-1} < t_{ij} \leq \gamma_{j, y_{ij}}) \right\}$$

$$\times \prod_{i \in D} \left\{ \frac{1}{\sqrt{b}} \exp\left[-\frac{(Z_i - a)^2}{2b} \right] \right\} \prod_{i \in D^c} \left\{ \exp\left(-\frac{Z_i^2}{2} \right) \right\}$$

$$\times \exp\left(\frac{-a^2}{18} \right) b^{-6} \exp\left(-\frac{5}{b} \right) \prod_{j=1}^{J} \left\{ (\sigma_j^2)^{-5} \exp\left(-\frac{5}{\sigma_j^2} \right) \right\}. \quad (5.17)$$

In (5.17), T_j denotes the set of images rated by rater j for the given treatment, and D and D^c denote the images belonging to the disease and disease-free groups.

Based on this expression, we can design an MCMC algorithm to sample from the posterior distribution in a manner similar to that employed earlier for the essay grades. Specifically, if we are given current values of $\mathbf{Z}$, σ^2, and $\mathbf{y}$, updates to the category cutoffs can be made exactly as in Cowles' algorithm. Likewise, the conditional distributions of $\{t_{ij}\}$ and σ^2, given all other parameters, are unchanged by the ROC assumptions, and so these parameters can also be updated using the conditional distributions described earlier.

The only conditional distributions affected by including a "disease" group in the model are the conditional distributions of $\mathbf{Z}$ and, of course, a and b, the parameters of interest.

The conditional distributions of the components of $\mathbf{Z}$ corresponding to the disease-free group are given in (5.6). From (5.17), it follows that the conditional distributions of the components of $\mathbf{Z}$ for the diseased group are also normally distributed. The mean and variance of Z_i are s/r and $1/r$, respectively, where r and s are given by

$$r = \frac{1}{b} + \sum_{j \in C_i} \frac{1}{\sigma_j^2} \qquad \text{and} \qquad s = \frac{a}{b} + \sum_{j \in C_i} \frac{t_{ij}}{\sigma_j^2}. \tag{5.18}$$

The conditional distribution of a is normally distributed with mean v/u and variance $1/u$, where

$$u = \frac{1}{9} + \sum_{i \in D} \frac{1}{b} \qquad \text{and} \qquad v = \sum_{i \in D} \frac{Z_i}{b}. \tag{5.19}$$

Finally, the conditional distribution of b given $\mathbf{Z}$ and a is an inverse-gamma distribution with parameters g and h, where g and h are given by

$$g = \frac{|D|}{2} + 5 \qquad \text{and} \qquad h = 5 + \frac{1}{2} \sum_{i \in D} (Z_i - a)^2,$$

and $|D|$ denotes the number of individuals in the disease group.

Simulation results

Using the conditional distributions specified above, an MCMC algorithm was designed to simulate from the posterior distributions of the parameters a and b for both the CB and PB SPECT collimators. For each sampled value of a and b, we computed the area under the corresponding ROC curve. Using the sampled areas from both treatments, we were thus able to directly compute the probability that the area of the CB ROC curve exceeded that of the PB ROC curve.

Histogram estimates of the ROC curve areas are depicted in Figure 5.9. These histograms were based on a MCMC run of 5,000 updates of all parameters, after first discarding 2,000 updates to ensure that the chain had reached equilibrium.

Several features of this plot merit comment. First, the maximum a posteriori estimates of the ROC areas for both PB and CB collimation are higher than the corresponding estimates for the standard analysis, reflecting the fact that these parameters represent the areas that would be obtained by an ideal judge. Also, the posterior distributions of these parameters are skewed, suggesting that p-values obtained under the assumption of asymptotic normality are likely to be inaccurate. The posterior probability that the area of CB exceeded that of PB was greater than 0.995. The classical p-value, although not interpretable as the probability that the

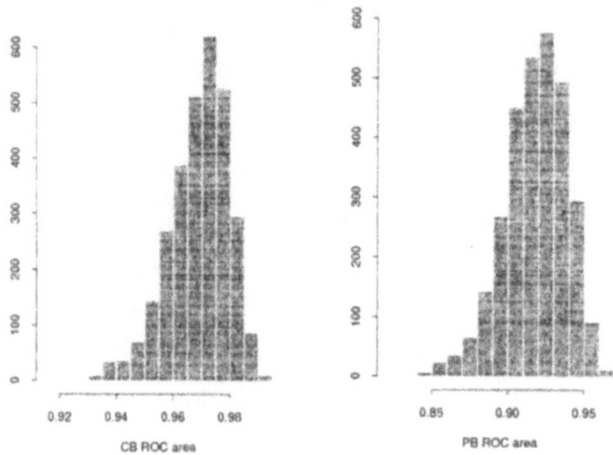

FIGURE 5.9. Histogram estimates of the ROC areas for CB and PB collimation.

area of CB is greater than the area of PB, was quoted as being less than 0.01 by Li et al. (1992).

5.5 Further reading

McCullagh and Nelder (1989) describe ordinal data models in which scale parameters of underlying latent variables are estimated as functions of covariates. By fitting the scale parameter as a function of the rater, the models in McCullagh and Nelder can be made to resemble the models described in this chapter, provided that appropriate constraints are used to ensure identifiability. Johnson (1996) describes multirater ordinal regression models for a similar set of essay grade data, but employs different forms of prior densities. Samejima (1969, 1972) describes graded response models that are similar to the multirater ordinal models described in this chapter, but from a classical perspective. Further ties between graded response models and multrater ordinal regression models are discussed in Chapter 7.

5.6 Exercises

1. The essay data described in this chapter is contained in the MATLAB mat-file essay.mat on the web site listed in the Preface. Reanalyze this data with the following questions in mind.

   a. Fit the multirater ordinal probit model without the third rater and without the regression of the essay traits on the grammatical variables.

b. Construct a normal scores plot of the latent residuals obtained for each rater. Are there raters whose latent residuals appear to violate the assumption of normality? How serious are these violations?

c. Plot the estimated latent essay traits of each rater against the latent essay traits obtained from the each of the others. Are there essays for which there is large disagreement among the raters?

d. Compare the estimates of the latent essay traits obtained without the third rater to the estimates obtained using data from all five raters. Did the third rater's data significantly change the essay rankings? How significant was this rater's data in determining the relative value of the rankings from the other raters; i.e., are the other rater variances robust to whether or not this rater's data was included?

2. In a survey conducted at Duke University, undergraduates were asked the following questions about an undergraduate statistics class:

Q1 Approximately how many hours per week did you spend on this course (including class meetings, discussion sessions, labs, reading and written assignments, exam preparation, projects, etc.)
    1. <2 hours. 2. 2–5 hours. 3. 5–9 hours. 4. 9–15 hours. 5. >15 hours.

Q2 What proportion of reading and written assignments did you complete?
    1. <50%. 2. 50–75%. 3. 75–85%. 4. 85–95%. 5. >95%.

Q3 What proportion of classes did you attend?
    1. <50%. 2. 50–75%. 3. 75–85%. 4. 85–95%. 5. >95%.

Q4 How difficult was the material taught in this class?
    1. Easy. 2. Less difficult than average. 3. Average. 4. More difficult than average. 5. Very difficult.

Q5 How much did you learn in this course compared to all courses that you have taken at Duke?
    1. Learned much less than average. 2. Learned less than average. 3. About average. 4. Learned more than average. 5. Learned much more than average.

Responses from this survey are contained in the file survey.dat from the web site listed in the Preface.

a. Fit a multirater ordinal probit model treating Q5 as the explanatory variable and Q1–Q4 as response variables. Did you treat Q5 as a continuous or factor variable? Why?

b. Construct normal scores plots of the posterior means of the latent residuals. Criticize the model fit.

c. For several combinations of covariate values, compute the fitted values for the response probabilities in Q1. Interpret the regression coefficients in language that a nonstatistician could understand.

6

Item Response Modeling

6.1 Introduction

In this chapter, we examine a model commonly used in educational testing, the item response model. In the educational setting, item response models are most often applied to tests containing questions that can be scored as being either correct or incorrect. Item response models are very similar to the multirater ordinal probit models described in the preceding chapter, although there are also important differences. The differences are reflected even in the terminology used to describe the models. In item response theory terminology, questions on a "test" are called *items*, and the individuals taking a test are called *examinees*. Generally, item response models do not incorporate regression relations of the type described in Chapter 5.

There are two primary objectives of item response theory. One is to assess the abilities of examinees. For example, a mathematics department might be interested in assessing the mathematics knowledge of incoming students, and so might administer a placement exam consisting of items involving basic algebra and trigonometry. The purpose of the test might then be to measure skills required for a first course in calculus.

The second major goal of item response modeling is to study the effectiveness of different test items in measuring an underlying trait—like algebraic or trigonometric aptitude. This aim is important because poorly designed tests cannot accurately measure examinee traits. For example, particularly difficult items are not useful in distinguishing students of low abilities, since almost all responses from such students are likely to be scored as incorrect. It is therefore important to match the difficulty of test items to the distribution of abilities in the examinee population.

In a mathematics placement exam given to college freshmen, questions involving Lie algebras probably do not serve a useful purpose.

Item response models are also common in sociological applications, as we illustrate below through the analysis of experimental data originally collected by Robert Terry. In this experiment, a group of 120 students interacted over a period of time, and 107 of the students then classified their fellow students in terms of their likeability, aggressiveness, and shyness. If, say, shyness is coded as a correct (present) or incorrect (absent) response for each student by each judge, then the structure of data from this experiment is very similar to testing data. Each of the 107 students who provided classifications of other student's shyness serves as an item, and the perceived shyness of each student functions as the examinee response.

In analyzing this study, we might be interested in the distribution of shyness for this group of students. Which students are generally perceived by their fellows as being shy, and which students tend not to be shy? We might also be interested in the items themselves. What factors influence a student's judgments about his or her peers? As judges of shyness, how do their definitions of shyness vary, causing them to differ in their selections of shy students? How precise are they in their definitions? Are they able to distinguish between small differences in personality, or are they relatively insensitive to such minor differences, assigning the label shy or not shy to a broad range of personalities?

6.2   Modeling the probability of a correct response

Let us begin our discussion of item response models (IRM) with a simple test in which one student provides a single response to a single item. The student can answer either correctly or incorrectly. As in case of binary data discussed in Chapter 3, we let y denote the student's response to the item, and take $y = 1$ if the response is correct and $y = 0$ if incorrect. We model the probability that the student responds correctly, $\Pr(y = 1)$, as a function of both the student's ability and the characteristics of the question on the test. As before, we employ a *latent trait* variable to represent the ability of each student, and introduce an *item response curve* to express the probability of that a student correctly answers a test item. The shape of the item response curve (IRC) depends on two parameters which indicate the difficulty of the item and the ability of the item to discriminate between students of low and high abilities.

6.2.1   Latent trait

We assume that an individual's performance on a particular item is dependent on an unknown characteristic θ, called a *latent trait* or *latent variable*. (In the literature of item response theory (IRT), latent variables are usually denoted by θ, rather than z or x as is more common in the statistical literature. For consistency with the IRT tradition, we adopt the former in this chapter.) In testing applications, latent traits represent student's knowledge of subject matter; in the sociological experiment, the latent trait represents the quality of shyness intrinsic to all students. In both cases, we assume that the latent trait is a continuous quantity which assumes values on the real line.

6.2.2   Item response curve

An item response curve models the probability that an individual answers an item correctly as a function of his or her latent trait θ. We assume that the item response curve takes the form of a distribution function belonging to a location-scale family. Denoting the canonical member of the location-scale family by $F(\cdot)$, an item response curve may thus be written in the form

$$\Pr(y = 1 \,|\, \theta) = F(a\theta - b).$$

Often, F is assumed to be a standard normal or standard logistic distribution function. The variables a and b are item-specific parameters which control the shape of the item response curve. We call a the slope, or *item discrimination*, parameter and b the intercept, or *item difficulty*, parameter.

A hypothetical item response curve is displayed in Figure 6.1. For this curve, we see that the probability of correct response is an increasing function of the latent trait. In almost all cases of practical interest, the item response curve is a monotonically increasing function because increased ability usually leads to an increased probability of correct response. On a mathematics placement exam, for example, the item response curve represents the probability of answering a mathematics item correctly as a function of mathematical ability. If an examinee is weak in mathematics—having a small value of θ—then the probability that the examinee correctly answers a question is expected to be small. In contrast, a student proficient in mathematics—having a large θ—is expected to correctly answer a mathematics question with high probability.

6.2.3   Item characteristics

Our description of an item response curve depends on two parameters, a and b. The second parameter, b, controls the location of the item response curve, or in substantive terms, the difficulty of an item. Figure 6.2 illustrates the effect of changing b for three item response curves having the same a. In this figure, F was taken to be a standard normal distribution, and the slope parameter a for all curves was 1. The values of b for the three curves are -1, 0, and 1. Note that all three curves are approximately parallel for values of θ close to zero and that the probability of a correct response for a student of given ability decreases as b changes from -1 to 1. If we center the distribution of student abilities so that $\theta = 0$ represents an average student, then an average student correctly answers a question with probability .15 for the item response with $b = -1$. Similarly, an average student answers the question with IRC parameter $b = 0$ with probability .5, and with probability .85 when $b = 1$. The item response curve drawn in Figure

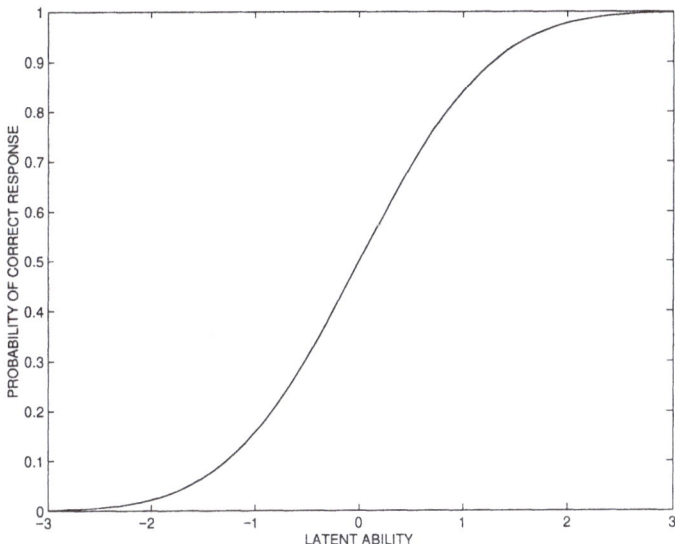

FIGURE 6.1. A typical item response curve.

6.2 corresponding to $b = 1$ represents an "easy" question on the test; the curve corresponding to $b = 0$ represents a question of moderate difficulty; and the curve with $b = -1$ represents a difficult question.

The slope parameter a of the item response curve indicates how well a particular item *discriminates* between students of varying abilities. Figure 6.3 illustrates the effect of changing the slope parameter a for three item response curves. As before, F is assumed to be a standard normal distribution, and the intercept parameter b was assigned a value of 0 for each of the curves in this plot. The slope parameter was assigned three values, $a = .5$, 1, and 2. As the slope parameter increases, the curve becomes steeper, reflecting the way in which the probability of a correct answer changes with examinee ability. For the curve with $a = .5$, the probability of a correct response changes only slightly as θ goes from 0 to 1. For such an item, the chance that a good student responds correctly is only slightly larger than the chance an average student does. Thus, this question does not discriminate well between average and good students. For the item response curve corresponding to $a = 2$, the probability of a correct response changes rapidly as θ increases from 0 to 1. A question with this item response curve does discriminate well between average and good students, and, thus, is useful to test administrators.

Figure 6.4 depicts item response curves for "ideal" and "useless" items. For discriminating above average and below average students, the item response curve with $b = 0$ and $a = \infty$ is ideal; the probability of a correct response on this item is 0 for any student with below average ability (θ less than zero) and 1 for any student with above average ability (θ greater than zero). An item having this item response curve is perfect in discriminating between below average and

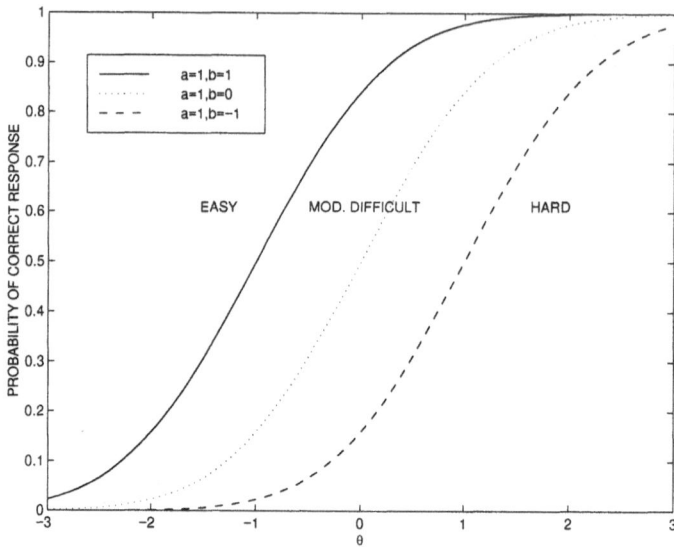

FIGURE 6.2. Three item response curves with the same discrimination but different difficulty levels. Each curve is labeled with its values of the item parameters *a* and *b*.

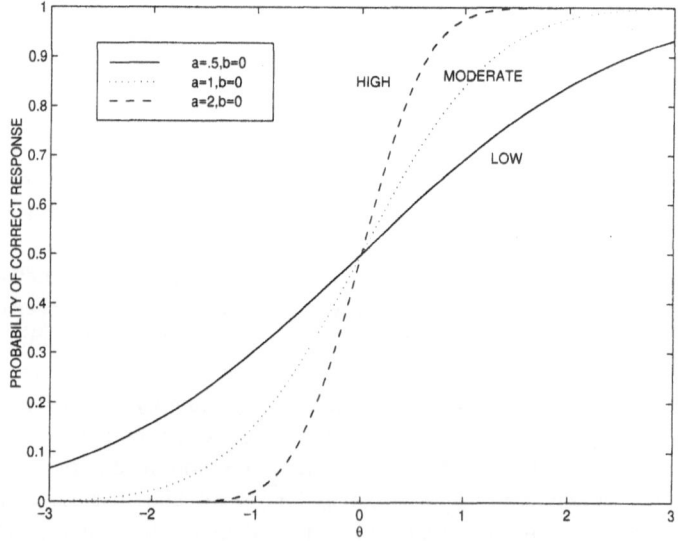

FIGURE 6.3. Three item response curves with the same difficulty and different discrimination levels. Each curve is labeled with the corresponding values of the item parameters *a* and *b*.

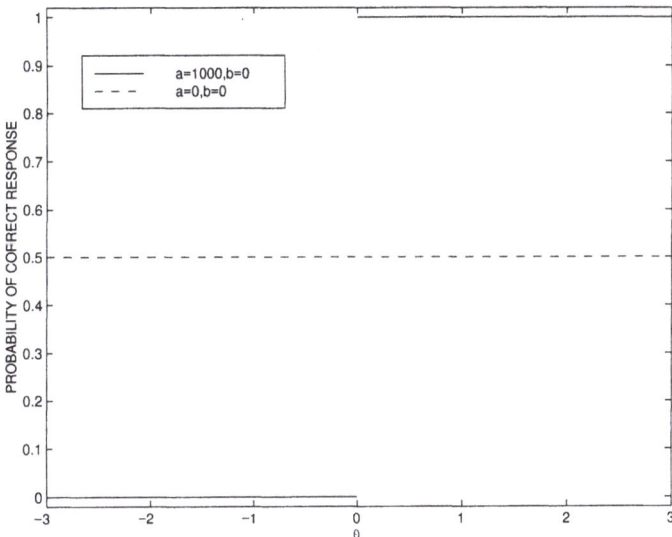

FIGURE 6.4. Item response curves corresponding to "ideal" and "useless" items. Each curve is labeled with the corresponding values of a and b.

above average students (but useless in discriminating between two above average students!). In contrast, the second item response curve corresponds to values of $a = 0$ and $b = 0$. All examinees, regardless of ability, have a 50% chance of correctly answering this item. Thus, this item offers no discrimination between students.

As an aside, it is worth contrasting the item response curve model formulation of this chapter to the latent trait formulation for ordinal data presented in previous chapters. To this end, consider again the multirater ordinal probit model described in Chapter 5. In that model, we assumed the existence of an underlying latent trait z_i that determined the probability that individual i would be classified into each of the available categories. If we further assume that individuals are classified on a binary scale ($K = 2$) by a judge whose variance parameter is σ_j^2 and whose upper category cutoff for a rating of 1 is γ_1, then the probability that this judge classifies an individual with trait z_i into category 2 is

$$\Pr(Y = 2) = 1 - \Phi\left(\frac{\gamma_1 - z_i}{\sigma_j}\right) = \Phi\left(\frac{z_i - \gamma_1}{\sigma_j}\right)$$

Taking $\Phi = F$ and interpreting $Y = 2$ as a "correct" response, we see that an item response model with probit link is nearly the same as the multirater ordinal probit model with $K = 2$. In terms of the parameters of the multirater ordinal probit model, $1/\sigma_j$ denotes the item discrimination parameter in the item response model, whereas γ_1/σ_j is analogous to the item difficulty parameter.

6.3   Modeling test results for a group of examinees

6.3.1   Data structure

So far, we have described two settings in which item response models might be applied. The first was a mathematics placement exam designed to measure precalculus mathematics ability of college freshmen. The second dealt with peer ratings. That data consisted of ratings from 107 student judges on the shyness of 120 of their peers. Each rated student was categorized as either "shy" or "not shy." Let us now describe the general way in which this type of data can be represented within the item response theory framework.

In both examples, there were n individuals (examinees) that were evaluated by a "test" containing k items. Each item was scored on a binary scale—either correct or incorrect, shy or not shy— so outcomes can be coded as 1 or 0. Thus, in both examples. the observed data can be represented by an $n \times k$ matrix of 1's and 0's. The first row of this matrix represents the sequence of responses recorded for the first examinee, the second row the sequence of responses received by the second examinee, and so on. Similarly, the first column contains the responses from all examinees to the first test item. If we let y_{ij} denote the response of the ith individual to the jth item on the test, the data derived from these experiments can be represented in the general form

$$
\begin{array}{c}
\begin{array}{cccc} \text{Item 1} & \text{Item 2} & \cdots & \text{Item } k \end{array} \\
\begin{array}{c} \text{Examinee 1} \\ \text{Examinee 2} \\ \vdots \\ \text{Examinee } n \end{array}
\begin{pmatrix}
y_{11} & y_{12} & \cdots & y_{1k} \\
y_{21} & y_{22} & \cdots & y_{2k} \\
\vdots & \vdots & \vdots & \vdots \\
y_{n1} & y_{n2} & \cdots & y_{nk}
\end{pmatrix}
\end{array}
$$

6.3.2   Model assumptions

We now develop a model for item response data using item response curves. As in the previous section, we assume that the ith individual's performance on a test depends on an unknown latent variable θ_i, and we let $\theta_1, \ldots, \theta_n$ denote the latent traits for all n individuals taking the test.

As a first assumption in our model, we assume that the probability that an individual answers a test item correctly depends only on their latent trait value and characteristics of the item. For the ith individual and jth item, we model this probability as

$$
\Pr(y_{ij} = 1 \mid \theta_i) = F(a_j\theta_i - b_j).
$$

As before, F represents a known distribution function and a_j and b_j are the item discrimination and item difficulty parameters associated with the jth item. It follows that the Bernoulli probability of response y_{ij} may be written

$$
\Pr(y_{ij} \mid \theta_i, a_j, b_j) = F(a_j\theta_i - b_j)^{y_{ij}}[1 - F(a_j\theta_i - b_j)]^{1-y_{ij}}, \qquad (6.1)
$$

for $y_{ij} = 0$ or 1.

The probability specified in (6.1) describes the probability that a single examinee correctly answers a single test item. To combine the results of this individual's responses on all items, we must make a further assumption: that of *local independence* . The term local independence in item response theory is similar to the usual definition of conditional independence in more common statistical parlance. In the present context, it simply means that the responses of an individual to items on the test are independent, given the latent trait value for that examinee and all item discrimination and difficulty parameters. From the assumption of local independence, responses by an individual to separate test items may be positively correlated, but this correlation must be entirely explained by the latent trait of the individual. (This fact has been exploited to test whether a single latent trait is adequate for modeling item response data; see, for example, Stout, 1990.)

Letting $y_{i1}, \ldots, y_{ik}$ denote the binary responses of the ith individual to k test items, and $\mathbf{a} = (a_1, ..., a_k)$ and $\mathbf{b} = (b_1, ..., b_k)$ the vectors of item discrimination and difficulty parameters for the test, respectively, from local independence we may express the probability of observing an individual's entire sequence of responses as

$$\Pr(y_{i1}, \ldots, y_{ik} \mid \theta_i, \mathbf{a}, \mathbf{b})$$
$$= \Pr(y_{i1} \mid \theta_i, \mathbf{a}, \mathbf{b}) \times \cdots \times \Pr(y_{ik} \mid \theta_i, \mathbf{a}, \mathbf{b})$$
$$= \prod_{j=1}^{k} F(a_j \theta_i - b_j)^{y_{ij}} [1 - F(a_j \theta_i - b_j)]^{1-y_{ij}}.$$

To combine the responses of all examinees, we also assume that the responses of the n individuals to the test items are independent. This assumption leads to a likelihood function that is proportional to the product of the probabilities of the sequences $y_{i1}, ..., y_{ik}$ over all examinees, or

$$L(\theta, \mathbf{a}, \mathbf{b}) = \prod_{i=1}^{n} \prod_{j=1}^{k} F(a_j \theta_i - b_j)^{y_{ij}} [1 - F(a_j \theta_i - b_j)]^{1-y_{ij}}. \qquad (6.2)$$

The total number of parameters contained in this likelihood function is thus $n + 2k$, assuming that the parametric form of F is known.

6.4   Classical estimation of item and ability parameters

In principle, maximum likelihood estimates of the item parameters $\{a_j\}$ and $\{b_j\}$ and the ability parameters $\{\theta_i\}$ can be obtained using standard numerical optimization procedures. Unfortunately, the estimates derived by jointly maximizing (6.2), called the *joint maximum likelihood estimates* or the joint MLE's, are problematic. For one thing, the parameters in this model are not identifiable, for reasons similar to those discussed in earlier chapters. To see this, consider again the model for the probability of a correct response:

$$\Pr(y_{ij} = 1 \mid \theta_i) = F(a_j\theta_i - b_j).$$

If each latent trait θ_i is multiplied by an arbitrary constant, say 2, and each item discrimination parameter a_j is divided by the same constant, then the predicted success probability is not changed. Thus, it is not possible to obtain unique estimates for both the latent abilities and item parameters. From a classical standpoint, several strategies for overcoming this problem of nonidentifiability have been proposed. One solution is to constrain the mean and variance of the estimated latent trait parameters to have a mean of 0 and standard deviation of 1. Another way to solve this problem is to place a probability distribution on the set of ability parameters; this was the approach used in the Bayesian models of the previous chapters, and it is the approach that we consider in what follows.

A second problem with joint maximum likelihood estimation is that the likelihood function may have multiple modes, and the mode obtained through numerical optimization procedures may not correspond to the mode yielding maximum likelihood globally. In addition, (joint) maximum likelihood estimation leads to estimates of parameter values that are often not reasonable from a practical perspective. To see why this might be so, recall that, in the sociological application, the item difficulty parameters $\{a_j\}$ represented the general tendency for judges to rate students as "shy." If one judge happened to rate every student as "not shy," then the maximum likelihood estimate for that judge's item difficulty parameter, $\hat{b}_j$, would be infinity, indicating that this judge would rate every imaginable student as shy. Such an estimate is probably not reasonable since we do not believe that this judge would actually rate all students in the population of interest as "not shy"—he or she would probably rate some students as shy if he or she rated a large enough group of students. A similar problem occurs if a judge rates all students as "shy."

Returning to the issue of identifiability, in many cases primary interest in an analysis focuses on inference for item parameters. When this is the case, *marginal maximum likelihood* may be used to establish identifiability of parameters and to estimate these parameters from data. This method is based on the idea of a marginal likelihood function that is obtained by integrating the ability parameters $\{\theta_1, ..., \theta_n\}$ out of the joint likelihood function:

$$L_m(\mathbf{a}, \mathbf{b}) = \int \cdots \int L(\theta, \mathbf{a}, \mathbf{b})h(\theta_1)\cdots h(\theta_n)\,d\theta_1\cdots d\theta_n. \qquad (6.3)$$

In this expression, the function $h(\theta) = \{h(\theta_1, \, dots, \theta_n)\}$ represents the assumed population distribution over the latent traits and can be given the same interpretation as a prior distribution. This function is sometimes estimated empirically using "prior" estimates of the item parameters (Bock and Aitkin, 1981), but it is more commonly is assumed to be a standard normal distribution.

Marginal maximum likelihood estimates, or MMLE's, are values of $\mathbf{a}$ and $\mathbf{b}$ that maximize the marginal likelihood function and were first proposed by Bock and Lieberman (1970). Computationally, finding the MMLE requires integrating the latent traits out of the likelihood function, and as Bock and Lieberman pointed out, this cannot be done easily using standard numerical algorithms. As a result, Bock and Lieberman's MMLE method was not commonly used until Bock and Aitkin (1981) described a variant of the EM algorithm that could be used to solve this numerical problem.

Generally, the MMLE is believed to have better sampling properties than the joint MLE. Typically, the marginal likelihood is unimodal, so the problem of convergence to a local maximum in the likelihood surface is eliminated. Also, the dimension of the MMLE is usually much smaller than the dimension of the joint MLE, and so asymptotic estimates of the standard errors of the MMLE tend to be closer to their actual sample variance. However, the MMLE cannot formally be applied in cases where the examinee traits are also required, unless the estimates of the examinee traits are based on conditional inference using the MMLE of the item parameters. Furthermore, the difficulty associated with items in which all examinees are rated either as successes or failures persists, and small sample properties of the MMLE cannot easily be studied.

6.5   Bayesian estimation of item parameters

To estimate item response model parameters using Bayesian methods, we must specify the prior distribution on all unknown parameters, including the ability parameters θ and item parameters $\mathbf{a}$ and $\mathbf{b}$. If the joint prior density on these parameters is denoted by $g(\theta, \mathbf{a}, \mathbf{b})$, then the posterior density is proportional to

$$g(\theta, \mathbf{a}, \mathbf{b}|\text{data}) \propto L(\theta, \mathbf{a}, \mathbf{b})g(\theta, \mathbf{a}, \mathbf{b}).$$

Of course, if interest focuses only on item parameters, then inferences concerning these parameters are based on the marginal posterior density for $\mathbf{a}$ and $\mathbf{b}$, found by integrating out the ability parameters from the joint posterior density:

$$g(\mathbf{a}, \mathbf{b}|\text{data}) = \int \cdots \int L(\theta, \mathbf{a}, \mathbf{b})g(\theta, \mathbf{a}, \mathbf{b})d\theta_1 \cdots d\theta_n.$$

Taking $g(\theta, \mathbf{a}, \mathbf{b}) = h(\theta)$, we see that this marginal posterior density is essentially the same density that is maximized to obtain the MMLE, which might more aptly be called the MMPE, or marginal maximum a posteriori estimate. The only difference between the two forms is the prior distribution on the item parameters; in the MMLE approach, this prior is assumed to be uniform; in the Bayesian framework, more realistic priors may be taken. Of course, inferences based on the MMLE use only the behavior of this density at its maximum value, whereas Bayesian inference relies on all of the information contained in this functional form.

6.5.1   Prior distributions on latent traits

One problem encountered in the classical item response model was non-identifiability of parameters. From within the Bayesian paradigm, this problem is eliminated by specifying a proper prior on the the distribution of latent traits, or, in other words, by assuming that the latent traits represent a random sample from a known population. This is equivalent to assuming that $\theta_1, ..., \theta_n$ are independently distributed from the probability distribution $h(\theta)$, where h represents the population of latent traits.

In this chapter, we follow standard practice and assume that h is a normal distribution with mean 0 and standard deviation 1. This prior distribution provides a useful scale for interpreting the values of estimated latent traits. An "average" individual has latent ability equal to 0, and essentially all latent traits are assumed a priori to fall in the interval $(-3, 3)$.

6.5.2   Prior distributions on item parameters

In general, there are likely to be two sources of prior information regarding the values of the item parameters. First, and perhaps less commonly, there may be prior information about specific values of the item parameters available from previous experiments. In the sociology application, for example, there may be related data collected on different personality traits, or other data collected from a different cohort of students. If such prior information is available, it should be incorporated using a suitable prior distribution.

Besides specific prior information concerning item parameters, it usually makes sense to specify prior distributions on the item parameters based on mathematical properties of the item response curve. For example, it is generally reasonable to assign high probability to positive values of the item discrimination parameters, since we expect that examinee responses to items are at least positively correlated with the latent traits they are intended to measure. Alternatively, if the latent trait distribution is assumed to be a standard normal distribution, it makes sense to place a small prior probability on values of the difficulty parameter outside of the range $(-5, 5)$, since for moderate values of the discrimination parameter, such items would seldom yield anything but all correct or all incorrect responses. Along similar lines, item discrimination parameters greater than 2 or 3, or less than 1/4 to 1/2, are often realistic, implying either that an item is extremely precise or almost useless in measuring the latent trait values.

To specify prior distributions on the item parameters, we seek distributions that reflect vague prior beliefs about the item parameters, but which regularize the posterior density and make fitting procedures more stable.

In performing the joint maximum likelihood estimation, we saw that item difficulty parameters became unbounded when an item is correctly (or incorrectly) answered by all examinees. To avoid this difficulty, we specify prior distributions on item parameters that discourage unusually small or large values. Independent normal distributions with mean 0 and standard deviation s_b, $s_b < 5$, serve this purpose. In other words, if there is an interval $(-B, B)$ in which, say, 95% of all the difficulty parameters are expected to fall, then we can represent this prior constraint by imposing a $N(0, s_b^2)$ on the item difficulty parameter by matching $2s_b$ to the upper limit of the desired interval, B.

Along similar lines, it is often useful to specify proper prior distributions to reflect vague beliefs about the values of the item discrimination parameters $\{a_j\}$. In practice, it is generally happens that item discrimination parameters corresponding to human judges are less than 3 when subjective judgments are made concerning the merit of individual rankings (assuming the population of traits follows a standard normal distribution). Furthermore, the correlation between a judge's ranking of an examinee and its underlying trait is usually—but not always—positive. This prior knowledge can be modeled by also specifying a normal distribution with mean $\mu_a > 0$ and standard deviation $s_a < \infty$ for the item discrimination parameters. The prior mean for a_j, μ_a, can usually be taken to be close to 1, indicating a moderate level of discrimination for items that are "average" in difficulty (see also Table 6.1). As in the case of the marginal maximum likelihood estimation method, the use of normal prior distributions on the item parameters also improves the stability of the fitting process. They also yield more reasonable estimates of the item parameters, even when anomalies exist in the data structure.

6.5.3 Posterior distributions

To summarize our prior beliefs, we assume a priori that

1. the latent abilities $\theta_1, \ldots, \theta_n$ are drawn from a standard normal distribution, and
2. the item slope parameters, $a_1, \ldots, a_k$, and intercept parameters, $b_1, \ldots, b_k$, are drawn from $N(\mu_a, s_a^2)$ and $N(0, s_b^2)$ distributions, respectively.

Combining this prior density with the likelihood, it follows that the posterior density is proportional to

$$g(\theta, \mathbf{a}, \mathbf{b}|\text{data}) \propto L(\theta, \mathbf{a}, \mathbf{b}) \prod_{i=1}^{n} \phi(\theta_i; 0, 1) \prod_{j=1}^{k} \phi(a_j; \mu_a, s_a^2)\phi(b_j; 0, s_b^2),$$

where $\phi(x; \mu, \sigma^2)$ denotes a normal density with mean μ and variance σ^2.

6.5.4 Describing item response models (probit link)

In many applications, primary interest focuses on summarizing the properties of test items. Although the characteristics of each item j are fully described by the two parameters a_j and b_j in the models considered here, it is often helpful to transform these parameters to a scale upon which they can be more easily interpreted. We describe convenient measures of the discrimination and difficulty characteristics of the items in the case where F is the standard normal distribution (the probit link).

One transformation useful in interpreting item response parameters results in the biserial correlation coefficient, defined as

$$r_j = \frac{a_j}{\sqrt{1 + a_j^2}}.$$

The biserial correlation measures the correlation between the responses $\{y_{ij}\}$ (with values 0 and 1) and the continuous latent traits $\{\theta_i\}$. It can be interpreted much like the standard correlation between two continuous variables. If $r_j = 0$, then the item and the latent trait are uncorrelated—the item provides no information about the underlying ability of the student. An item which discriminates well between individuals of different abilities will have a large value of a_j and a correlation r_j that is close to 1.

In the preceding discussion, we saw that the intercept parameter b_j determined the location of the item response curve, which, in turn, determined the difficulty level of the jth item. As an alternative measure of item difficulty, we can consider the response y_{ij} of a randomly selected examinee to the jth test item. The probability that a randomly selected examinee correctly answers an item is obtained by averaging over the latent trait distribution of examinees, or

$$p_j = \int \Pr(y_j = 1|\theta)h(\theta)d\theta. \tag{6.4}$$

The quantity p_j can be interpreted as the proportion of students in the population that will correctly answer the jth item. In the case that the distribution of latent traits follows a standard normal distribution, it can be calculated as

$$p_j = \Phi\left(\frac{-b_j}{\sqrt{1 + a_j^2}}\right). \tag{6.5}$$

To illustrate the relationships between the biserial correlation and probability of correct response, and the item discrimination and difficulty parameters, we have listed the values of each parameter for the six curves displayed in Figures 6.2 and 6.3. The table also contains qualitative descriptions of these curves in terms of difficulty and discrimination. The values for a_j and b_j displayed in the table are typical of slope and intercept parameters observed in many sociological and testing applications.

6.6   Estimation of model parameters (probit link)

The marginal posterior distributions of item response model parameters cannot be obtained analytically, so once again we must resort to MCMC methods to summarize our findings. In the case that the link function F is assumed to be a standard normal distribution, the MCMC algorithm of Section 3.3 can be modified in a straightforward way to simulate values from the posterior distribution of interest.

Curve	a_j	b_j	r_j	p_j	Difficulty	Discrimination
1	.5	0	.45	.5	moderate	low
2	1	0	.71	.5	moderate	moderate
3	2	0	.89	.5	moderate	high
4	1	−1	.71	.76	high	moderate
5	1	0	.71	.50	moderate	moderate
6	1	1	.71	.24	low	moderate

TABLE 6.1. Biserial correlation r_j and probability of a correct response p_j for six item response curves with slope a_j and intercept b_j. Qualitative descriptions of the levels of difficulty and discrimination for these curves are also given.

As in the binary regression models of Chapter 3, we suppose that there exists a continuous latent variable that underlies each binary response. Specifically, suppose that corresponding to each binary observation y_{ij}, there exists a continuous unobservable variable Z_{ij} which is normally distributed with mean $m_{ij} = a_j\theta_i - b_j$ and standard deviation 1. A success ($y_{ij} = 1$) is observed if the latent variable Z_{ij} exceeds 0, and a failure ($y_{ij} = 0$) results when the latent variable Z_{ij} is less than 0.

With the introduction of the continuous latent data $\mathbf{Z} = (Z_{11}, ...Z_{nk})$, the item response model assumptions can be reexpressed in the following way:

- Latent values $\{Z_{ij}\}$ are assumed to be drawn independently from normal distributions with means $m_{ij} = a_j\theta_i - b_j$ and standard deviation 1.
- The observed data $\{y_{ij}\}$ are indicators of the values of $\{Z_{ij}\}$; that is, $y_{ij} = 1$ if Z_{ij} is positive, and $y_{ij} = 0$ if Z_{ij} is negative.
- A priori, the latent traits $\{\theta_i\}$ are assumed to be drawn independently from a normal distribution with mean 0 and standard deviation 1.
- A priori, the slope parameters $\{a_j\}$ are independently distributed according to a normal distribution with mean μ_a and standard deviation s_a. The intercept parameters $\{b_j\}$ are independently distributed from a normal distribution with mean 0 and standard deviation s_b.

It follows that the joint posterior density of all model parameters is given by

$$g(\mathbf{Z}, \theta, \mathbf{a}, \mathbf{b}|data) \propto \prod_{i=1}^{n}\prod_{j=1}^{k} \left[\phi(Z_{ij}, m_{ij}, 1)\text{Ind}(Z_{ij}, y_{ij}) \right]$$

$$\times \prod_{i=1}^{n} \phi(\theta_i; 0, 1) \prod_{j=1}^{k} \left[\phi(a_j; \mu_a, s_a^2)\phi(b_j; 0, s_b^2) \right],$$

where $\text{Ind}(c, d)$ is equal to 1 when $\{c > 0, d = 1\}$ or $\{c < 0, d = 0\}$, and equal to 0 otherwise.

6.6.1   A Gibbs sampling algorithm

A Gibbs sampling procedure can be used to sample from the joint posterior distribution over the entire collection of unknown parameters and latent data. The sampler

is based on iteratively drawing values from three sets of conditional probability
distributions:

- $g(\mathbf{Z}|\theta, \mathbf{a}, \mathbf{b}, \text{data})$,
- $g(\theta|\mathbf{Z}, \mathbf{a}, \mathbf{b}, \text{data})$,
- $g(\mathbf{a}, \mathbf{b}|\mathbf{Z}, \theta, \text{data})$.

To implement the Gibbs sampler, suppose at iteration $t-1$ the current values of
the model parameters are denoted by $\{Z_{ij}^{(t-1)}\}$, $\{\theta_i^{(t-1)}\}$, $\{a_j^{(t-1)}\}$, and $\{b_j^{(t-1)}\}$. Then
one complete cycle of the Gibbs sampler can be described as follows:

1. Values of the latent data $\{Z_{ij}\}$ are simulated conditionally on the current values
   of the latent traits and item parameters . The conditional posterior distribution
   of Z_{ij} is a truncated normal distribution with mean $m_{ij} = a_j^{(t-1)}\theta_i^{(t-1)} - b_j^{(t-1)}$
   and standard deviation 1. The truncation of this posterior distribution depends
   on the value of the corresponding binary observation y_{ij}. If the observation is
   a success (1), the truncation of Z_{ij} is from the left at 0; if the observation is a
   failure (0), the truncation is from the right at 0. Let the (new) latent data value
   simulated from this truncated normal distribution be denoted by $\{Z_{ij}^{(t)}\}$.
2. Latent traits $\{\theta_i\}$ are simulated from their posterior distribution conditionally
   on current values of the latent data and item parameters. Using the latent data
   representation, the item response model can be written as

   $$Z_{ij}^{(t)} + b_j^{(t-1)} = a_j^{(t-1)}\theta_i + e_{ij},$$

   where the error terms e_{ij} are independent normal with mean 0 and standard
   deviation 1. For a given value of i, this is a special case of the familiar normal
   linear model with unknown parameter θ_i. Combining the above sampling model
   with the $N(0, 1)$ prior, it follows that the conditional posterior density of θ_i is
   normally distributed with mean

   $$m_{\theta_i} = \frac{\sum_{j=1}^{k} a_j^{(t-1)}(Z_{ij}^{(t)} + b_j^{(t-1)})}{\sum_{j=1}^{k} a_j^{2(t-1)} + 1}$$

   and variance

   $$v_{\theta_i} = \frac{1}{\sum_{j=1}^{k} a_j^{2(t-1)} + 1}.$$

   Let $\theta_i^{(t)}$ denote a random draw from this normal density.
3. The item parameters $\{a_j, b_j\}$ are simulated from their joint posterior density
   conditionally on the current values of the latent data and the latent traits. To
   determine this conditional distribution, rewrite the latent data model as

   $$Z_{ij}^{(t)} = a_j\theta_i^{(t)} - b_j + e_{ij}.$$

   Since the values of the latent data $\{Z_{ij}\}$ and the latent traits $\{\theta_i\}$ are fixed, this
   (for a fixed value of j) can be viewed as a normal linear model with unknown
   parameter vector (a_j, b_j) and known covariate vector $(\theta_i^{(t)}, -1)$. Let X^* denote

the design matrix for this regression, having columns $(\theta_i^{(t)}, -1)$, and let Σ_0 denote the prior covariance matrix

$$\Sigma_0 = \begin{bmatrix} s_a^2 & 0 \\ 0 & s_b^2 \end{bmatrix}.$$

It follows that the conditional posterior density of $\{a_j, b_j\}$ is also multivariate normal with mean vector

$$m_j = \left[X^{*'} X^* + \Sigma_0^{-1} \right]^{-1} \left[X^{*'} Z^t + \Sigma_0^{-1} \begin{pmatrix} \mu_a \\ 0 \end{pmatrix} \right]$$

and covariance matrix

$$v_j = (X^{*'} X^* + \Sigma_0^{-1})^{-1}.$$

Let $(a_j^{(t)}, b_j^{(t)})$ denote a value sampled from this distribution.

Given suitable starting values for the parameter values, these steps define one iteration in a Gibbs sampling scheme that can be used to obtain samples from the posterior distribution over all model parameters. Convergence of the algorithm is typically obtained within several hundred observations and is typically not very sensitive to the choice of starting values. Of course, good initial values certainly do not hurt.

One strategy for obtaining reasonable initial values for model parameters is to first set all θ_i's to 0, based on the fact that their prior mean is 0. Next, initialize all item discrimination parameters to μ_a, again in accordance with their prior means. Using this initial estimate of the discrimination parameters $\{a_j\}$, the item difficulty parameters $\{b_j\}$ can be initialized by solving for b_j in the equation for the probability of a correct response (6.5). Thus, an initial values for $\{b_j\}$ are be obtained from

$$b_j^* = -\sqrt{1 + \mu_a}\, \Phi^{-1}(\hat{p}_j),$$

where $\Phi^{-1}()$ denotes the standard normal quantile function and $\hat{p}_j$ denotes the proportion of individuals responding correctly to item j.

6.7   An example

As an illustration of this methodology, consider the analysis of the "shyness" data described earlier. One hundred seven judges each rated individual students in a class of size 120 with respect to the variable "shyness." An observed value of $y_{ij} = 1$ indicates that the jth judge rated the ith student as "shy," and $y_{ij} = 0$ indicates a student was judged "not shy." The data matrix of the $\{y_{ij}\}$ consists of 120 rows and 107 columns—a given row of the matrix corresponds to the ratings collected from all judges for a particular student.

Based on the criteria discussed above, we take independent $N(1, 1)$ priors for components of $\mathbf{a}$, and independent $N(0, 1)$ priors for the components of $\mathbf{b}$. Using

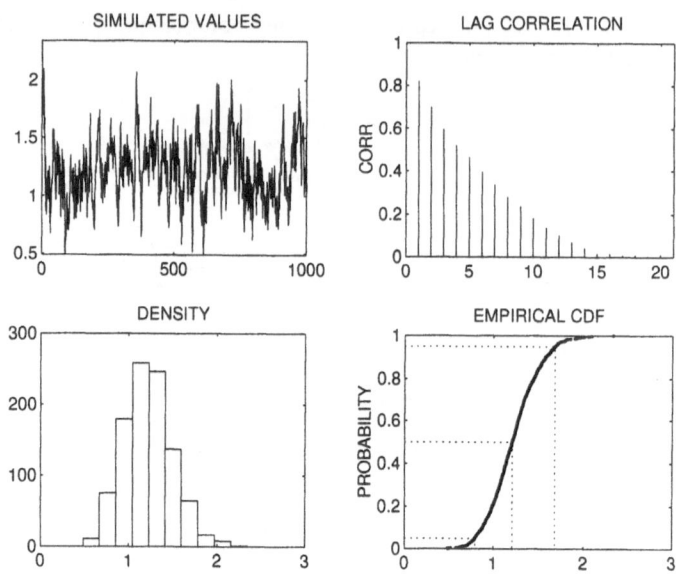

FIGURE 6.5. Convergence graphs for the simulated values of a_{25}. The top left graph displays the sequence of simulated values and the top right graph displays the lag correlations of the sequence as a function of of the lag value. The bottom left graph is a histogram of the simulated values and the bottom right graph is the empirical distribution function.

this model specification, we ran the Gibbs sampler for a total of 1000 iterations to obtain estimates of the distribution on the model parameters.

The output of the MCMC simulation was a set of 1000 simulated values of the item parameters $\{(a_j, b_j), j = 1, ..., 107\}$, as well as estimated values of the latent trait variables $\{\theta_i\}$.

Before using this set of simulated values to summarize the posterior distribution, we must decide if the MCMC algorithm had converged to the posterior density before it was terminated. One simple way of addressing this question is to examine plots of simulated parameter values. For example, Figure 6.5 displays several convergence plots for an arbitrarily chosen item discrimination parameter, a_{25}.

The top left graph depicts the sequence of simulated values of a_{25} against the iteration number. For this particular parameter, there appears to be no general trend with iteration number, either in the mean of the series or in its variance, suggesting that the 1000 iterations obtained from the sampler might be adequate for burn-in. However, the values in the plot seem to be highly correlated.

The correlation between successive iterates is quantified in the lag correlation plot in the upper right panel of Figure 6.5. This plot depicts the autocorrelation of the sequence of simulated values as a function of the lag time. Note that the lag 1 correlation of the sequence of simulated values is .8, and that the lag correlations are close to 0 by lag 15.

The bottom plots summarize the distribution of the simulated values of a_{25}. The left panel shows a histogram of the sampled values, and the right depicts their empirical distribution function. From the histogram, we see that the posterior density of this parameter was approximately normally distributed and concentrated most of its mass between .5 and 2. From the empirical cdf, we find that the locations of the median, and 5th and 95th percentiles were approximately 1.2, 0.8, and 1.7, respectively.

Although it appears that the MCMC algorithm had burned-in by iteration 1000, there was evidence of moderate autocorrelation in the simulated values. This suggests that a larger sample might be needed to accurately summarize the posterior distribution. To assess the simulation error in this sample, the method of batch means described in Section 2.6 was used. Based on this method, we estimated that the posterior mean of a_{25} was 1.22, and that there was an approximate simulation standard error of .03. Because .03 represents only a small fraction of the posterior uncertainty in this parameter value, we might be satisfied in using the initial run of 1000 iterates to summarize our model, particularly if similar simulation errors were observed for other model parameters. This turned out to be the case; therefore the simulated values from this relatively short run of the Gibbs sampler were used to summarize our posterior beliefs about the characteristics of different judges in this experiment.

6.7.1   Posterior inference

Point estimates of the item response curves can be plotted using the posterior means of the item parameters, which can also be used to compare the performance of different judges by simply plotting these parameters against one another. Such a plot is displayed in Figure 6.6; each point is labeled by the number of the corresponding judge. It is interesting that all posterior means of the difficulty parameters $\{b_j\}$ lie between -2 and 2. Most of the posterior means of the discrimination parameters $\{a_j\}$ fall between 0 and 2.

As mentioned above, the discrimination and difficulty characteristics of the judges can be viewed in terms of the biserial correlation parameters $\{r_j\}$ and the marginal "probabilities of being shy" $\{p_j\}$. Using the simulated values of a_j and b_j, the posterior distributions of these parameters can be estimated from their empirical distributions. For example, the posterior mean of r_{25}, say, can be computed by first computing the value of $r_{25} = a_{25}/\sqrt{1 + a_{25}^2}$ for all simulated values of a_{25}, and then taking the sample mean of these values. Figure 6.7 displays the posterior means of $\{r_j\}$ and $\{p_j\}$ calculated in this way.

From Figure 6.7, we see that there is large variation in the posterior means of the p_j across judges. This indicates that the judges varied greatly in their general perception of shyness among the students. A majority of judges had marginal probability values between .15 and .7; however, five judges displayed values of p_j close to 1, and each of these judges rated most of the students as shy. We also see considerable variability in the posterior means of the correlation parameters

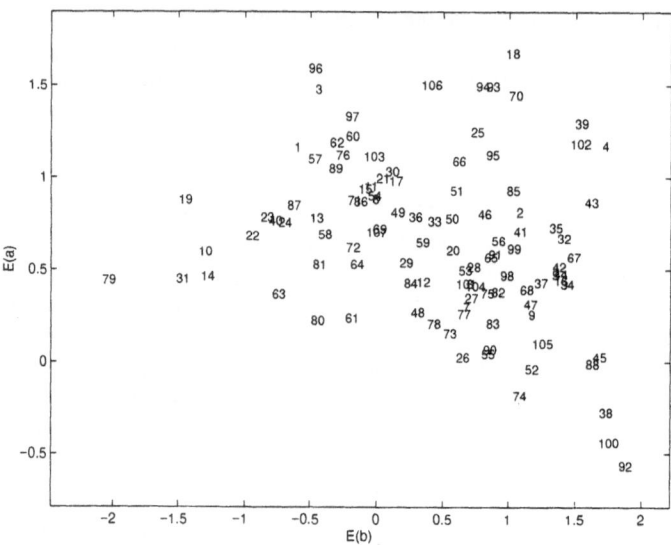

FIGURE 6.6. Plot of the posterior means of $\{b_j\}$ against the posterior means of $\{a_j\}$. Each point is labeled with the number of the corresponding judge.

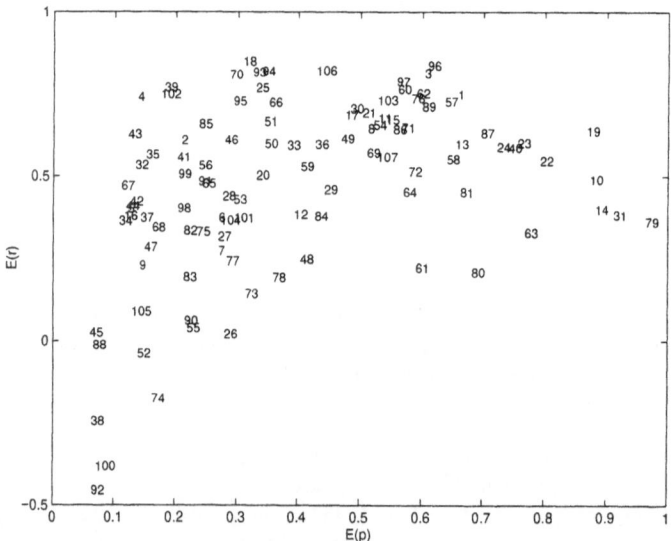

FIGURE 6.7. Graph of the posterior means of the probabilities of correct response $\{p_j\}$ against the posterior means of the biserial correlations $\{r_j\}$. Each point is labeled with the number of the corresponding judge.

r_j. Most judges have correlation values in the range (.2, .8), although several were estimated as having negative values.

There is an interesting relationship between a judge's rating of examinees, as measured by p_j, and his or her ability to discriminate between different levels of shyness, as measured by the parameter r_j. Judges who had small average values of p_j also tended to have small average values of r_j. As the values of the posterior mean of p_j increased, the posterior means of r_j also increased, up to about the value $p_j = .5$, and then slowly decreased. One interpretation of this pattern is that the judges that rated approximately half of the students as shy were the best discriminators of students with different levels of shyness.

Figure 6.7 displays only the posterior means of the item response curve parameters, but it says nothing about the variability of the posterior densities of these parameters. For each parameter, a simulated sample is available and can be used to summarize the variability of the item response curves and derived quantities. Figure 6.8 shows one such plot for the posterior densities of the slope parameters $\{a_j\}$. In this figure, the 90% credible regions are depicted along with the estimated posterior mean of each parameter.

This plot might be used for deciding which judges were useful in discriminating between students of different levels of shyness. If a particular judge's error bar was located entirely above the horizontal line at $a = 0$, then the probability that this judge's discrimination parameter was positive was greater than .9. On the other hand, error bars that cover or fall below 0 correspond to judges who probably cannot accurately assess the characteristic of shyness within their peer group.

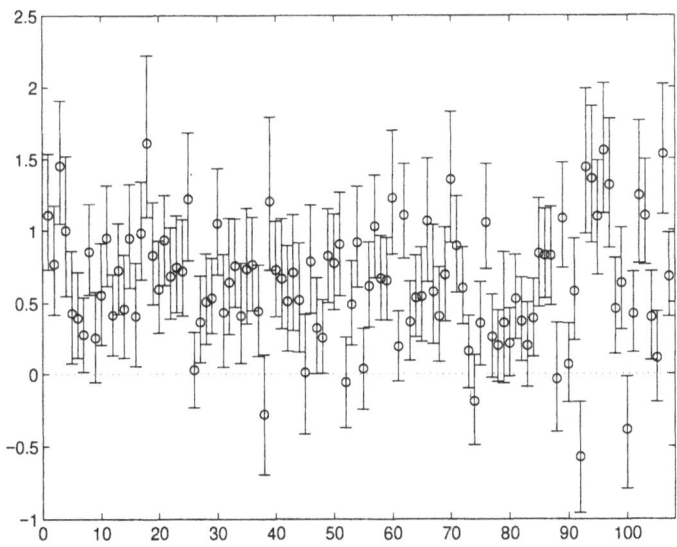

FIGURE 6.8. Graph of posterior densities of the slope parameters $\{a_j\}$ for all judges. The center of a error bar corresponds to the posterior mean and the extremes correspond to the 5th and 95th percentiles of the posterior density.

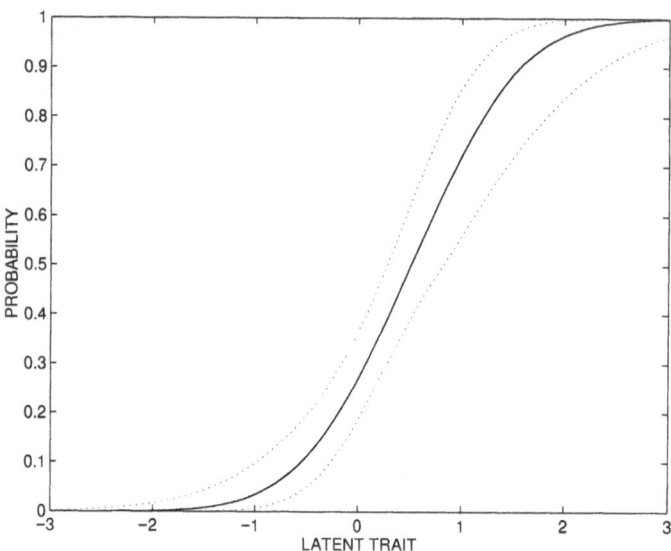

FIGURE 6.9. Plot of posterior density of the item response curve for judge 25. The solid line corresponds to the location of the posterior medians and the dashed lines correspond to the 5th and 95th percentiles of the posterior density.

Simulated values of the item response parameters can also be used to study the shape of the item response curves for particular judges. Figure 6.9 depicts the posterior density of the item response curve for judge 25. To construct this plot, values of the probability $p_j(\theta) = \Phi(a_j\theta - b_j)$ were simulated over a range of θ values between -3 and 3. The solid line in the figure represents the median of these values for each value of θ, and the dashed lines depict the 5th and 95th percentiles of the item response curve. A student with average shyness corresponds to a θ value of 0. Looking at the figure for this value of θ, we see that judge 25 rated "averagely shy" students as shy with probability .3, on average. The 90% credible region for this probability extends from approximately .2 to .4.

6.8   One-parameter (item response) models

The item response model discussed above assumes that k test items are adequately described by two parameters $\{a_j\}$ and $\{b_j\}$, representing the item's discrimination and difficulty level. For obvious reasons, this model is referred to as the *two-parameter item response model* since there are two parameters associated with each item on the test. Occasionally, this model is simplified by assuming that the discrimination levels across items are equal. In a testing scenario, this means that each item has the same ability to discriminate between examinees. Assuming that all of the slope parameters are equal to one, we are left with the *one-parameter item response model,*

$$\Pr(y_{ij} = 1 \mid \theta_i) = F(\theta_i - b_j).$$

6.8.1   The Rasch model

If in addition to fixing all of the discrimination parameters at a common value, we also assume that the link distribution is a standard logistic distribution function, we obtain the well-known *Rasch model*

$$\Pr(y_{ij} = 1 \mid \theta_i) = \frac{\exp\{\theta_i - b_j\}}{1 + \exp\{\theta_i - b_j\}}.$$

The Rasch model has a number of desirable properties which make it attractive in item response modeling. For one, its simple form makes it easy to interpret. The family of Rasch item response curves resemble the item response curves graphed in Figure 6.2, having the same basic shape and differing only in respect to their location. Another convenient property of the Rasch model is that the estimation procedure for the difficulty parameters $\{b_j\}$ does not depend on the ability parameters $\{\theta_i\}$. To see this, note that given responses $\{y_{ij}\}$, the likelihood function for the Rasch model may be written

$$L(\mathbf{b}, \theta) = \prod_{i=1}^{n} \prod_{j=1}^{k} \left(\frac{\exp\{\theta_i - b_j\}}{1 + \exp\{\theta_i - b_j\}} \right)^{y_{ij}} \left(1 - \frac{\exp\{\theta_i - b_j\}}{1 + \exp\{\theta_i - b_j\}} \right)^{1 - y_{ij}}$$

$$= \frac{\prod_{i=1}^{n} \exp\{y_{i.}\theta_i - \sum_{j=1}^{k} b_j y_{ij}\}}{\prod_{i=1}^{n} \prod_{j=1}^{k} (1 + \exp\{\theta_i - b_j\})},$$

where $y_{i.}$ is the total number of correct responses for the ith individual, $\mathbf{b}$ is the vector of difficulty parameters, and θ is the vector of ability parameters. Examining this expression, we see that the subject totals $\{y_{i.}\}$ are sufficient statistics for the ability parameters $\{\theta_i\}$ (conditionally on the item parameters $\{b_j\}$). Furthermore, it can be shown that the conditional distribution of the responses $\{y_{ij}\}$, given the subject totals, does not depend on the ability parameters. This fact has motivated a *conditional maximum likelihood estimation* procedure in which the difficulty parameters $\{b_j\}$ are maximized conditionally on subject totals. The conditional MLE's so defined are much easier to compute than the joint MLE's or marginal MLE's described earlier in this chapter, and so have found common use.

6.8.2   A Bayesian fit of the probit one-parameter model

The Rasch model can, of course, also be interpreted from within the Bayesian framework. The two primary differences between the Rasch model and the two-parameter item response model discussed in Sections 6.1-6.7 are use of a logit rather than probit link, and the constraint that the item discrimination parameters be assigned a common fixed value, i.e., that $a_j = 1$ for all j.

The constraint that all item discrimination parameters be assigned a fixed value of 1 causes little difficulty in either the Bayesian model or its associated estimation procedures. Assuming that a probit link is still used to model the item response curve, the parameters in the constrained model can be estimated just as they were in the case of the two-parameter model using the Gibbs sampler described in Section 6.6.1. The only modification of this sampler occurs in step (3), where item parameters are updated. If a one-parameter model is assumed, then item difficulty parameters are not sampled, but are instead maintained at their initial values.

In terms of parameter estimation, modeling item response curves using a logistic distribution function instead of a normal distribution function is a bit more complicated. Because the logistic distribution function does not possess the convenient sampling properties of the normal distribution function, the full-conditional distributions required for the Gibbs sampler of Section 6.6.1 are not available when a logistic link is employed. Instead, Metropolis-Hastings steps must be used to update parameter values. Of course, in practice there is little difference between the predictions obtained using the two models, and so the probit link might be chosen to facilitate computations.

6.9   Three-parameter item response models

In the previous section, we described how the two-parameter item response model could be reduced to a one-parameter model by assuming that the discrimination parameters for all items were equal. Moving toward the other extreme, it is sometimes necessary to expand the two-parameter model to a three-parameter model. This extension is most often considered in the context of multiple-choice tests, in which students of every ability have a non-neglible probability of answering an item correctly simply by guessing. The two-parameter item response model implicitly assumes that students of very low ability have probability 0 of answering a question correctly; the three-parameter model corrects this model deficiency by incorporating a guessing parameter.

Like the two-parameter model, the three-parameter model uses an item discrimination parameter a_j and an item difficulty parameter b_j to model the response probabilities for each item. In addition, a third item parameter, c_j, called the *guessing parameter* is introduced to represent the lower limit for the probability of a correct response for a student of arbitrarily low ability. With this additional parameter, the probability that an examinee correctly answers an item is represented as the sum of the probabilities that the examinee guesses and gets the item correct (c_j), plus the probability that the examinee does not guess $(1 - c_j)$ and gets the item correct ($F(a_j\theta_i - b_j)$); that is,

$$\Pr(y_{ij} = 1|\theta_i) = c_j + (1 - c_j)F(a_j\theta_i - b_j).$$

Simulation from posterior distributions on three-parameter item response models that utilize probit links ($F = \Phi$) can be performed using a generalization of the Gibbs sampling algorithm described in Section 6.6. The generalization requires the introduction of independent latent Bernolli random variables $\{u_{ij}\}$, where u_{ij} has success probability c_j and represents the event that student i correctly guessed the answer for item j. The observed response y_{ij} is related to the latent variables Z_{ij} and u_{ij} according to

$$y_{ij} = u_{ij} + (1 - u_{ij}) \times I(Z_{ij} > 0).$$

Values of the latent data $\mathbf{Z}$, the parameters $\mathbf{a}$, and $\mathbf{b}$, and the abilities $\{\theta_i\}$ are simulated from distributions similar to those given in Section 6.6. The guessing parameters $\{c_j\}$ are simulated from independent beta densities with parameters that depend on the guessing outcomes $\{u_{ij}\}$. Details of this fitting algorithm are provided in Sahu (1997).

6.10 Model checking

6.10.1 Bayesian residuals

Two classes of item response models, the one-parameter and two-parameter models, have been introduced. Generally speaking, two-parameter models provide better fits to actual data, but regardless of which model is entertained, it is important to examine model fit. As in the case of ordinal regression models, assessing model fit requires an examination of residuals.

Residuals in item response models can be defined in the usual way. If $p_{ij} = F(a_j\theta_i - b_j)$ denotes the probability that individual i responds correctly to the jth item, then one definition of a residual is the difference between the observed binary value of individual i's response to item j and the corresponding probability of a correct response,

$$r_{ij} = y_{ij} - p_{ij}.$$

From a Bayesian perspective, the value of the fitted probability p_{ij} is obtained from the posterior probability on θ_i, a_j, and b_j, while the observed value of y_{ij} is assumed to be constant.

As in the case of binary regression, it is difficult to assess the magnitude of r_{ij} because of the binary nature of the response variable. To cope with this difficulty, we can either use the diagnostics developed for binary regression in Chapter 3, or alternatively, because the number of examinees in item response models tends to be large, we can group individual responses together according to latent ability. The grouped responses are less discrete and can be more easily examined through diagnostic plots. Suppose then that we group the responses for a particular item, say item j, into L groups using the posterior distribution on the ability parameters $\theta_{1j}, ..., \theta_{nj}$. Each group is defined by an interval on the latent ability scale, and we let the centers of these intervals be denoted by $\theta_1^{(g)}, ..., \theta_L^{(g)}$. For the lth interval of ability values, let $\hat{p}_{lj}^{(g)}$ denote the proportion of individuals in that class that correctly answered the jth item.

A convenient way to display grouped residual values is to plot the estimated item response curve for the jth item against the midpoints of the ability intervals $\{\theta_l^{(g)}\}$ and the observed proportions $\{\hat{p}_{lj}^{(g)}\}$.

By comparing the observed item response curve with the posterior distribution of the fitted item response curve, we can judge if the model provides a reasonable fit to the data. To focus on the differences between the data and the fit, we can also plot the posterior distributions of the residuals $r_{lj} = \hat{p}_{lj}^{(g)} - p_j(\theta_l^{(g)})$ for all intervals $l = 1, ..., L$. Model lack of fit is indicated by intervals in which the posterior distribution of the grouped residuals are concentrated away from zero.

6.10.2 Example

To illustrate residual analyses in item response models, we now examine the fit of the two-parameter item response model to the shyness data. For brevity, we confine our analysis to the ratings of judges 18, 19, 21, and 55.

As a first step in the analysis, we grouped the responses from the 120 students on these items into 11 intervals based on the posterior means of student's latent abilities. Next, we plotted the observed item response curve for each judge; these plots are presented on the left panel of Figure 6.10. The posterior median, along with the 5th and 95th% percentile points, for the item response curves are also provided.

Several interesting features are apparent from these plots. First, it is clear from the graphs that these four judges exhibited very different item response curves. The observed item response curve for judge 18 shows a sharp increase between the values $\theta = 0$ and 1, indicating good discrimination between individuals of low and high shyness. In contrast, the item response curve for judge 19 is close to 1 over a broad range of θ, indicating both low discrimination for shyness and the fact that this judge rated nearly all students as shy. Judge 55's observed curve is nearly horizontal; this particular judge was not able to distinguish between individuals of different levels of shyness.

From Figure 6.10, we also see that for all of these judges, the posterior distributions of the fitted probabilities seem to agree well with the observed item response curves. This is not unexpected since the two-parameter model includes a difficulty and discrimination parameter for each item. The adequacy of this fit is reflected in the residual posterior distributions shown on the right side of Figure 6.10. Most of the credible regions for the residuals contain zero, with only a few concentrating their mass on regions not containing zero.

For purposes of comparison, we also fit the one-parameter probit model to the same data. The observed and fitted item response curves for this model are displayed on the left side of Figure 6.11. Recall that in this model, we assume that the discrimination abilities of the 107 judges are identical; this fact is reflected by the similar slope of the item response curves for the four judges. From Figure 6.11, we see that the fit of this model is not as good as it was for the two-parameter model. For example, the observed probabilities for judge 18's evaluations are

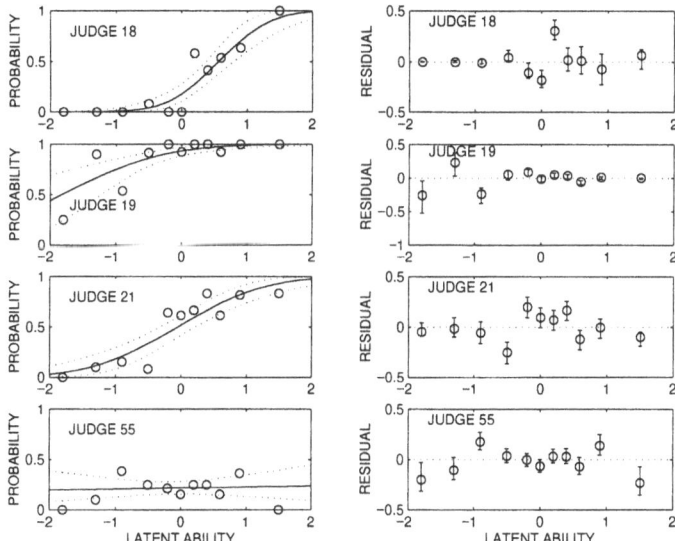

FIGURE 6.10. Investigation of model fit for the two-parameter model for the shyness dataset. The left panel depicts observed and fitted probability distributions for four judges; the right illustrates the associated residual posterior distributions for the same judges.

smaller than the corresponding fitted probabilities for negative values of θ but tend to exceed the fitted probabilities for positive θ. The exact opposite of these trends are exhibited in the residual plots for the 55th judge, and there also appears to be a relatively large residual present in the residual plot of the 19th item. Based on the graphical evidence provided by Figure 6.11, we conclude that the two-parameter model better describes this dataset.

6.11   An exchangeable model

6.11.1   Prior belief of exchangeability

In the preceding discussion, we considered two models for describing peer ratings of student shyness: the one-parameter and two-parameter item response models. The two-parameter model assumed that each judge had a distinct discrimination ability, quantified by a_j. In contrast, the one-parameter model assumed that all judges were equally discriminating, and included a single discrimination parameter a. In fitting of the two-parameter model, we found that many of the estimated discrimination abilities were similar in size, suggesting that the simpler one-parameter model might be adequate for representing the observed data. However, in our residual analysis of the one-parameter model, we noticed significant lack of fit for several judges who displayed unusually high or low discrimination abilities.

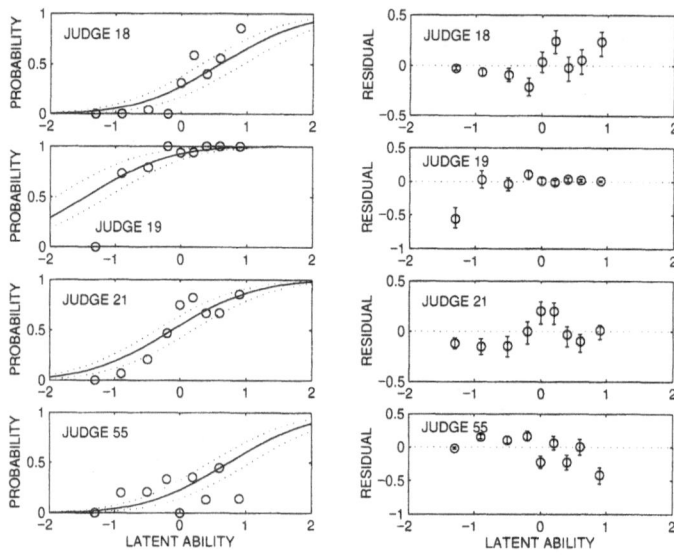

FIGURE 6.11. Investigation of model fit for the one-parameter model for shyness dataset. The left panel shows the observed and fitted probability distributions for four randomly selected judges. The right panel depicts the residual posterior distributions for the same judges.

As a compromise between the one- and two-parameter models, we might consider an *exchangeable* item response model in which we shrink item discrimination parameters toward a common value. From a practical standpoint, this allows us to use the responses obtained from one judge to improve the estimation of another judge's discrimination parameter, and it also provides a convenient mechanism for summarizing the distribution of discrimination parameters within the rater population of interest.

To motivate an exchangeable model for the discrimination parameters, suppose that, a priori, we are uncertain about the magnitudes of the individual item discrimination parameters, but we believe that all items are likely to be approximately equal for discriminating individuals of high and low abilities. In other words, we wish to model the fact that the value of the discrimination parameter for one judge is probably similar to the values of the discrimination parameters for other judges. Such knowledge is best described using an exchangeable prior model on the discrimination parameters.

Exchangeability of item discrimination parameters can be incorporated into the two-parameter model by means of a hierarchical prior distribution. This hierarchical model is constructed in a fashion similar to that employed in Chapter 1 for modeling exchangeability of binomial proportions.

A hierarchical prior distribution is specified in two (or more) stages. In the first stage, we assume that the parameters $a_1, ..., a_k$ are independently drawn from a common distribution which depends on several unknown parameters. The parameters of this distribution are called hyperparameters. In the second stage of the model, we place a prior distribution on the values of the hyperparameters from the first stage. Typically, the prior specified over the hyperparameters is vague. In this way, we model the fact that the first stage parameters $(a_1, \ldots, a_k)$ are identically distributed a priori, but that we are uncertain as to precisely what their distribution is.

Because the item discrimination parameters are real-valued, it is convenient to use a normal model in the first-stage of the hierarchy. Thus, we assume that the item discrimination parameters $a_1, \ldots, a_k$ are independently and identically distributed according to a normal distribution with unknown mean μ and standard deviation τ. In the second stage of the model, we assign vague priors to the hyperparameters μ and τ. We take a locally uniform prior for μ and an inverse-gamma density with parameters v_1 and v_2 for τ^2. With these assumptions, the joint prior density for the vector of slope parameters $\mathbf{a}$ and hyperparameters (μ, τ^2) can be written

$$g(\mathbf{a}, \mu, \tau^2) = (\tau^2)^{-v_1 - 1} \exp\{-v_2/\tau^2\} \prod_{j=1}^{k} \phi(a_j; \mu, \tau^2). \qquad (6.6)$$

The parameters of the first-stage normal prior on the discrimination parameters have clear interpretations: μ represents the mean of the discrimination parameters, and τ^2 models the variation between the discrimination parameters. If it is known that individual judges rate students with approximately equal discrimination, a prior distribution for τ^2 that concentrates its mass near zero should be specified. Of course, the value $\tau = 0$ corresponds to a one-parameter item response model since in that case, all item discrimination parameters are assumed to be equal. However, we assume a vague prior distribution on the value of τ^2, indicating that we do not know how similar the item discrimination parameters will be. Specifically, we take $v_1 = 1$ and $v_2 = 1$.

6.11.2   Application of a hierarchical prior model to the shyness data

Combining the assumptions of the two-parameter item response model with the hierarchical model of the last section, we obtain the following joint posterior density.

$$g(\mathbf{Z}, \theta, \mathbf{a}, \mathbf{b}, \mu, \tau^2 | \text{data}) \propto \prod_{i=1}^{n} \prod_{j=1}^{k} [\phi(Z_{ij}, m_{ij}, 1) I^*(Z_{ij}, y_{ij})]$$

$$\times \prod_{i=1}^{n} \phi(\theta_i; 0, 1) \prod_{j=1}^{k} \phi(b_j; 0, s_b^2) g(\mathbf{a}, \mu, \tau^2),$$

Here, $g(\mathbf{a}, \mu, \tau^2)$ is the function obtained by taking $v_1 = v_2 = 1$ in (6.6).

To obtain posterior samples from this model when applied to the shyness dataset, we modified the Gibbs sampler of Section 6.6.1 to reflect the changes in the full-conditional distributions that resulted from the introduction of the hierarchical prior. More specifically, in step (3) of that algorithm, s_a^2 (in the prior covariance matrix) and μ_a were replaced with μ and τ^2 to reflect changes in the conditional distributions of the discrimination parameters. Because the full-conditional distributions of the latent data and latent traits were not changed by the introduction of the second-stage prior, steps (1) and (2) of the sampler remained the same. Finally, two new steps were added. In the first, μ was updated using a normal density with mean equal to the sample mean of all item discrimination parameters a_j and variance τ^2/k. In the second, τ^2 was updated using an inverse-gamma distribution with shape parameter $k/2 + \nu_1$ and scale parameter $\nu_2 + \sum_j (a_j - \mu)^2/2$.

To demonstrate the effect of the exchangeable prior on the posterior estimates of the discrimination parameters, in Figure 6.12 we plotted the posterior means of these parameters under the exchangeable model against the corresponding estimates obtained from the two-parameter model. As expected, the hierarchical prior had the effect of shrinking the discrimination parameters toward a common mean value of about .7. As stated earlier, this shrinkage represents a compromise between the unrestricted two-parameter model and the (infinitely) restricted one-parameter model.

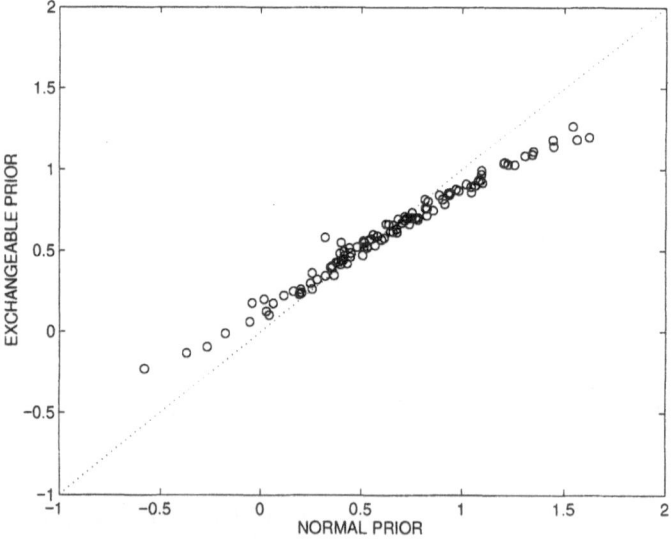

FIGURE 6.12. Scatterplot of posterior means of the item slope parameters with a normal prior against the posterior means using an exchangeable prior.

6.12   Further reading

Although this chapter has focused on the Bayesian fitting of item response models, there is a broad literature on the classical fitting and description of this class of models. Hambleton and Swaminathan (1985) and Baker (1992) present general reviews of classical methods, and van der Linden and Hambleton (1997) illustrate the generalization of item response theory for categorical responses. Bock and Aitkin (1981) illustrate the use of the EM algorithm for computing maximum likelihood estimates of the item parameters.

Early Bayesian analyses of item response models are found in Swaminathan and Gifford (1982, 1985) and Tsutakawa and Lin (1986). Albert (1992) and Bradlow, Wainer, and Wang (1999) illustrate the use of Gibbs sampling for modeling probit item response models and Patz and Junker (1997) demonstrate the use of Metropolis within Gibbs simulation algorithms for fitting Bayesian item response models with a logistic link.

6.13   Exercises

1. Consider the two-parameter item response model where the distribution function F is the standard logistic distribution. The probability of a correct response on a single item is given by

$$\Pr(y = 1|\theta) = \frac{\exp\{a\theta - b\}}{1 + \exp\{a\theta - b\}}.$$

   where a and b are item-specific parameters.

   a. Plot the item response curves (for values of θ between -3 and 3) for the cases $(a, b) = (1, 0)$ and $(a, b) = (1, 1)$. Describe the effect of changing the parameter b on the location and shape of the item response curve.
   b. Plot the item response curves for the cases $(a, b) = (1, 0)$ and $(a, b) = (2, 0)$. Describe the effect of changing the parameter a on the location and shape of the item response curve.
   c. Give values of the item parameters a and b which correspond to an easy item in which everyone gets a correct response. Plot the item response curve for this item.
   d. Give values of the item parameters a and b which correspond to an item which is ineffective in discriminating between individuals of different abilities. Plot the item response curve for this item.

2. At Bowling Green State University, all entering freshmen are required to take a mathematics placement test. This test is used to assess the high school math background of the students and recommend an appropriate first college math class. Form B of this test consists of 35 multiple-choice questions on topics in high school algebra. The dataset place98.dat contains the results of the placement exam for 200 BGSU freshmen. The data matrix $\{y_{ij}\}$ consists of 200

rows and 35 columns, where rows index the 200 students taking the test and columns the 35 questions. The entries of the matrix are 0's and 1's, where a 1 in the ith row and jth column indicates that the ith student got the jth question correct.

The difficulty of the jth item can be estimated by the proportion of students responding correctly, say $\tilde{p}_j$. The discrimination ability of the jth item can be estimated by computing the Pearson correlation between the binary responses $\{y_{1j}, ..., y_{nj}\}$ and the total scores $\{t_1, ..., t_n\}$, where $t_i = \sum_{j=1}^{k} y_{ij}$— call this correlation r_j. Compute the observed proportion $\tilde{p}_j$ and the correlation r_j for all 35 questions. Draw histograms of the proportion values and the correlation values. Plot the pairs $\{(\tilde{p}_j, r_j)\}$ as a scatterplot. Comment on the differences in difficulty level and discrimination levels of the 35 items. Are there items that appear to be particularly difficult or easy? Are there items that appear to discriminate poorly or very well among students of different abilities?

3. (Mathematics placement test continued)
Fit a two-parameter probit item response model of the form

$$\Pr(y_{ij} = 1) = \Phi(a_j\theta_i - b_j)$$

to the mathematics placement dataset using the Gibbs sampling algorithm described in the text. For prior distributions, assume that the slope parameters $\{a_j\}$ are independently distributed from a normal distribution with mean 1 and standard deviation 2. Assume also that the intercept parameters $\{b_j\}$ are drawn independently from a normal distribution with mean 0 and standard deviation 2. Run the algorithm for 2000 iterations.

  a. To assess convergence of the MCMC run, graph the simulated values of each slope parameter a_j against the iteration number. Also, construct a lag correlation plot for each sequence of simulated values (as in Figure 6.5). Comment on the magnitude of autocorrelation evident from the graphs.
  b. Compute the posterior mean of each of the 35 slope parameters $a_1, ..., a_{35}$. Using the batch means algorithm described in Chapter 2.5, compute the simulation standard error for each posterior mean estimate.
  c. Compute $P(a_j > 2|y)$ for $j = 1, ..., 35$. Using the batch means algorithm, compute the simulation standard error for each posterior probability estimate.
  d. Rerun the Gibbs sampler to obtain a new simulation sample of size 2000. Compute the posterior means of all of the slope parameters and compare your estimates with the values computed in the first simulation run. Also, recompute the posterior probabilities in part (c). Comment on the differences between the estimates found in the two simulation runs. Should a larger simulation sample be taken?

4. (Mathematics placement test continued) In this problem, it is assumed that the two-parameter probit item response model specified in Exercise 3 has been fit. Focus here lies on different summaries of the posterior distribution.

a. Plot the posterior means of the $\{b_j\}$ against the posterior means of the $\{a_j\}$. Compare this graph with the scatterplot of the $\{(\tilde{p}_j, r_j)\}$ that was drawn in Exercise 2.

b. Plot the posterior mean of the item response curve (like Figure 6.9) as a function of θ for three different questions on the test.

c. One way of assessing if the discrimination parameter a_j is significantly larger than 0 is to compute the posterior probability $P(a_j > 0|y)$. Compute these probabilities for all items and graph these probabilities as a function of the item number. Which items appear to be effective discriminators among poor, average, and good students?

d. Consider a student with "above-average" ability, $\theta = 1$. For this student, consider the probability that he or she answers the jth question correct,

$$p_j = \Pr(y_{ij} = 1) = \Phi(a_j - b_j).$$

For a selected question on the math placement test, say question k, construct a histogram for the probability that this student answers correctly. Using the simulated values, find a 90% probability interval for p_k.

e. For the same "above-average" student with $\theta = 1$, find the probability that he or she has a better than even chance of getting the first question correct (i.e. $\Pr(p_1 > .5)$). Compute this probability for the remaining 34 questions on the test.

5. (Mathematics placement test continued) In the fit of the two-parameter probit item response model, it was assumed that the a_j were, a priori, independently drawn from normal distributions with means 1 and standard deviations 2, and the b_j were independently drawn from $N(0, 2)$ distributions. To examine the effects of these prior assumptions on the marginal posterior distribution of the item parameters, consider the alternative choices for these normal priors. Alternative priors are more vague if large values, like 5, are assigned to the normal standard deviations. Alternatively, the priors can be made more precise by specifying small values of the standard deviations. For each new prior specification, rerun the Gibbs sampling algorithm and compute the posterior means of the a_j and the posterior probabilities that $a_j > 1$. Compare your answers with those found in Exercise 4. Did the choice of prior distribution matter in this case?

6. (Mathematics placement test continued) As an alternative to the two-parameter probit model described in Exercise 3, fit the one-parameter probit model

$$\Pr(y_{ij} = 1) = \Phi(\theta_i - b_j)$$

to the math placement dataset using the Gibbs sampling algorithm. Assume that the intercept parameters are a priori distributed from a normal density with mean 0 and standard deviation 2. Repeat parts (b), (d), and (e) of Exercise 3. Comments on any differences between inferences made using the one- and two-parameter models.

7. (Mathematics placement test continued) Use the procedure described in Section 6.9 to assess the goodness-of-fit of the one-parameter probit model to the math placement dataset.

   a. Choose one question that appears to discriminate between students of different abilities.
   b. Group the individual students into ten classes by their estimated ability.
   c. For each ability class, compute the proportion of students who answered the item correct. Plot these proportions against the class midpoints to get an observed item response curve for this question.
   d. Find the posterior mean of the item response curve for this question. Plot this curve on top of the observed item response curve obtained in part (c).
   e. By comparing the fitted item response curve with the observed curve, comment on any systematic differences between the two curves. These differences indicate potential lack of fit of the one parameter model.
   f. Repeat the procedure described in parts (a)–(e) for a second question that appears to have a low discriminatory ability.

8. (Mathematics placement test continued) Consider the application of the hierarchical prior model on the math placement dataset. Suppose that the 35 questions on the test have similar discrimination by assuming the following two-stage prior:

   - $a_1, ..., a_{35}$ are independent $N(\mu, \tau^2)$
   - μ and τ^2 are independent, μ is assigned a vague flat prior, and τ^2 is assigned an inverse-gamma prior with parameters v_1 and v_2. Prior information on τ^2 can be assumed to be relatively vague by taking $v_1 = v_2 = 1$.

   Assume that the item slopes $\{a_j\}$ and the item intercepts $\{b_j\}$ are independent. As in Exercise 3, assume that the b_j are independently drawn from $N(0, 4)$ distributions.

   a. Fit the implied hierarchical model using 2000 iterations of Gibbs sampling algorithm described in Section 6.10.
   b. From the simulation output, compute the posterior means and posterior standard deviations of the item slope parameters $\{a_j\}$. Compare the values of the posterior means with the posterior means from the fit of the regular two-parameter item response model in Exercise 3. Show (perhaps by a graph such as Figure 6.12) how the posterior means using the exchangeable prior shrink the usual parameter estimates toward a central value.

7
Graded Response Models: A Case Study of Undergraduate Grade Data

In this chapter, we discuss the extension of item response models—in which items are scored as being either correct or incorrect—to graded response and partial credit models, in which items are classified into ordered categories. To put this extension into proper perspective, it is useful to first examine the relationship of item response models to binary regression models and multirater ordinal regression models.

As we found in the last chapter, responses of examinees to a single item on a test can be studied using item response models. However, such data can also be modeled using standard binary regression models as described in Chapter 3. In the case of binary regression models, examinee ability parameters play the role of a single covariate in the regression, and with only one item, the item discrimination parameter can be assigned the value 1, since in that case, comparisons with other items are not relevant. Indeed, if a logistic link function is used in the binary regression model, then the corresponding item response theory model is simply a Rasch model; by introducing an indicator variable representing each item, the equivalence of the Rasch model to a logistic regression model extends to tests with an arbitrary number of questions. Graphically, the close relationship between item response models and binary regression models can be illustrated by comparing Figure 3.2 to Figure 6.1. The shaded areas in the subplots of Figure 3.2 represent the predicted probabilities that two individuals with latent abilities of 1.21 (top) and -0.43 pass their statistics class. If shaded areas under similar curves for students of arbitrary latent ability are plotted against latent ability, an item response curve similar to that depicted in Figure 6.1 is obtained.

In the more general setting of multiple-item tests where discrimination parameters for different items are allowed to vary, the simple binary regression models of Chapter 3 cannot be adapted to obtain the item response models described in Chapter 6. Instead, the multirater ordinal modeling framework of Chapter 5 is required. The connection between multirater ordinal regression models and item response models can be understood by comparing a multirater ordinal regression models used to model data containing two possible response categories to item response models in which all discrimination parameters are required to be positive. Defining the raters in the multirater ordinal regression models to be the items in the item response model and letting the examinee traits represent the individuals rated in the multirater ordinal model, we find that the likelihood functions (see Section 6.2.3) for these models are equivalent. Graphically, item response curves can be obtained from the curves in Figure 5.3 by again plotting the area under the curves and to the right of 0 as a function of latent ability. In this case, the curve corresponding to Rater 1 would have a higher discrimination parameter than the curve for Rater 2, since Rater 1's variance parameter is smaller.

Graded response models are the natural extension of item response models for data recorded in more than two ordered categories. For example, graded response models might be applied to test data in which partial credit is assigned to items. In this case, examinee scores on test items might take values from, say, 1 to C, where C corresponds to an item being entirely correct. Based on the relation of item response models to multirater ordinal data models, the preceding discussion suggests that graded response models can be considered as a special case of multirater ordinal regression. Statistically, the two models have essentially the same form.

Graded response models were first proposed by Samejima (1969, 1972) for the analysis of multiple-item tests in which examinee responses were graded on an ordinal scale. She also investigated the theoretical properties of estimators used in graded response models and provided consistency and uniqueness results of marginal maximum likelihood estimates. Related to graded response models are partial credit models (e.g., Masters, 1982). Partial credit models are special cases of graded response models in which all item response curves are assumed to be a location-translated logistic distributions. As mentioned above, both partial credit and graded response models can be considered special cases of the multirater ordinal regression models as discussed in Chapter 5; the connections between these models is discussed more fully by Thissen and Steinberg (1984) and Mellenberg (1994). The recent text *Handbook of Modern Item Response Theory* (1997) by van der Linden and Hambleton discusses over 20 extensions of these model forms to various psychometric settings.

Because the basic modeling issues for graded response models were described in Chapter 5 under a slightly different guise, in this chapter we simply illustrate their application to the important and difficult problem of analyzing undergraduate grade data. This application has previously been discussed by Johnson (1997) and is interesting for several reasons. First, the definition of the latent trait or latent ability in this setting is more subtle than it is in most graded response model applications. This subtlety leads naturally to a discussion of higher-dimensional traits. In addition, because undergraduate grade records play such an important role in our educational system, numerous approaches for analyzing these data have been proposed and it is interesting to compare the results obtained using graded response models (both Bayesian and non-Bayesian) to results obtained under alternative model assumptions.

7.1 Background

College grades, like elementary and secondary school grades, pervade our educational system and affect everyone involved in it. In fact, nearly all college-educated Americans know what their GPA was (or is), and to some extent measure, their academic success by this number.

Unfortunately, the use of GPA as a measure of academic success can be very misleading and lies at the heart of numerous problems experienced at many undergraduate institutions. Problems associated with GPA arise when grades obtained from different instructors in different courses are averaged as if they all had the same interpretation. This practice creates systematic biases against students enrolled in more rigorous curricula and has important consequences in the student's course selection. It creates perverse incentives for faculty to inflate grades and lower standards and it rewards students for selecting less challenging courses and majors (Larkey and Caulkin, 1992).

To understand the problems caused by the use of GPA, consider the data illustrated in Figure 7.1. This figure contains boxplots of classroom mean grades for all undergraduate classes with enrollments of 20 or more students offered at Duke University in 12 selected departments between the fall semester of 1989 and spring semester of 1994. Two aspects of the grade assignment process are clear from these plots. First, there is substantial variation in the median grades assigned in different departments. Second, there are even greater differences in the grading patterns between instructors within the same departments. As a consequence of these differences, students taking a majority of their classes in, say, Department 1 are likely to finish college with higher GPAs than students who take a majority of their classes in Department 2. In fact, the comparative advantage for Department 1 students may be even greater than indicated by Figure 7.1 because there is evidence that departments with high-ability students tend to grade more stringently than those with lower-ability students (Goldman and Widawski 1976).

Such differences in grading patterns have grave implications for our educational system. Besides the obvious inequities inflicted upon students enrolled in "hard" majors, differences in grade distributions result in a substantial reduction in the number of courses taken by students in subjects like mathematics and the natural sciences, as well as other challenging upper-level undergraduate courses. In fact, Larkey and Caulkin (1992) estimated that several hundred thousand fewer mathematics and natural sciences courses may be taken each year in the United States as a direct result of differential grading policies.

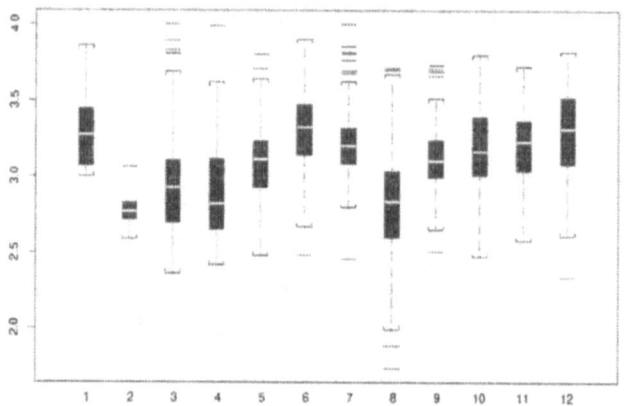

FIGURE 7.1. Boxplots of mean classroom grades assigned in classes offered by 12 Duke University departments. These 12 departments coincide with the departments studied in Goldman et al. (1974): cultural anthropology, biochemistry, biology, chemistry, economics, engineering, history, mathematics, political science, psychology, sociology, and Spanish. For reasons of confidentiality, the departments were arbitrarily relabeled with the integers 1 through 12.

7.1.1   Previously proposed methods for grade adjustment

To address the problem of differential grading criteria, numerous alternative measures for student performance have been proposed in the educational literature. Several of the more prominent are presented below.

Pairwise-comparison methods

Goldman and Widawski (1976) proposed a grade adjustment method based on pairwise comparisons of grades obtained by the same students across multiple departments. In their method, the difference in a student's grades for classes taken in different departments provides information about the relative grading standards between the departments. Goldman and Widawski averaged all such differences obtained from the transcripts of 475 University of California at Riverside (UCR) students to obtain grade adjustment factors for 17 academic fields. Based on this analysis, Goldman and Widawski concluded that there were systematic differences in grading patterns across academic departments at UCR and that departments with high-ability students tended to grade more stringently than fields with less able students.

Goldman and Widawski's analysis was extended by Strenta and Elliott (1987) and Elliott and Strenta (1988) in studies of Dartmouth College undergraduates. In Strenta and Elliot (1987), pairwise comparisons of departments were restricted to introductory courses, and external measures of student abilities (SAT and high school GPA) were used for validation. The results corroborate the earlier conclusions of Goldman and Widawski and confirm a stable trend in differential department grading standards over a 10-year period and between public and private institutions. In their later article, Elliott and Strenta (1988) incorporated both within- and between-department course comparisons, and estimated grade adjustments for a larger number of departments.

Graded response models

In his 1989 doctoral thesis, Young adapted a graded response model for application to undergraduate grade data. In his model, the grade assigned assigned to student i in class j was denoted by y_{ij}, and the underlying variable representing the ith student's ability was denoted by θ_i. The item discrimination parameter (inverse of the instructor variance parameter) for class j was denoted by a_j, and b_{jk} denoted the upper grade cutoff for grade k in class j. Assuming that a total of C grades were possible, the basic assumption of Young's graded response model (GRM) model was that the probalility that student i received a grade of at least k in class j was equal to

$$\Pr[y_{ij} \le k] = \pi_{ijk} = \frac{\exp[a_j\theta_i - b_{jk}]}{1 + \exp[a_j\theta_i - b_{jk}]}. \tag{7.1}$$

Note that this expression represents a logisitic multirater ordinal regression model in which y_{ij} denotes the data, $1/a_j$ denotes the rater variance parameter, θ_i the latent ability of the i individual, and b_{jk} the upper cutoff for a grade of k in class j.

Young applied this model to a cohort of Stanford undergraduate grades and found that estimates of student abilities obtained using this model correlated better with external measures of student abilities than did raw GPA (Young 1990, 1993). For example, the multiple correlation of student abilities obtained from the GRM model with verbal and mathematics SAT scores and high school GPA was higher than it was for raw GPA.

Regression models

Larkey and colleagues at Carnegie Mellon University investigated linear techniques for adjusting student GPAs to account for the difficulty of courses taken (Caulkin, Larkey, and Wei, 1996; Larkey and Caulkin, 1992; see also Young, 1992). In the simplest and perhaps most useful version of their approach, an additive adjustment was made to each student's GPA based on estimates of the difficulty of the student's curriculum. The difficulty of courses was estimated from a linear regression of student grades on "true" student GPA and course difficulty parameters. If y_{ij} again denotes the grade of the ith student in the jth class, and g_i and c_j denote the ith student's true GPA and difficulty of the jth class, respectively, then the additive model used to estimate these adjustment factors takes the form

$$y_{ij} = g_i - c_j + e_{ij}. \tag{7.2}$$

In (7.2), e_{ij} represents a mean zero, normally distributed error term. Numerical procedures for estimating model parameters are described in Caulkin, Larkey, and Wei (1996).

Caulkin, Larkey, and Wei report that predictions obtained using the additive adjustment model produced estimates of student performance that correlated more highly with high school GPA and SAT scores than did estimates obtained using Young's GRM. The performance of the additive adjustment model on a cohort of Duke University students is examined further in Section 7.4.2.

7.2   A Bayesian graded response model

Each of the grade adjustment methods described attempt to simultaneously account for the two critical factors that affect grade assignment: the achievement levels of students within a class and instructor-specific grade cutoffs relative to these perceived achievement levels. We now describe a Bayesian version of a graded response model that incorporates each of these ideas in a coherent statistical framework.

7.2.1   Achievement indices and grade cutoffs

The Bayesian graded response model is premised on the assumption that an instructor assigns grades by first ordering perceived student performance from best to worst, possibly with ties for groups of students who performed at approximately the same level. This ranking is assumed to be based on an achievement index that is implicitly defined by the instructor. The particular definition of achievement for each class is arbitrary.

After ordering students according to their estimated classroom achievement, instructors next group students into grade categories by fixing grade cutoffs between the estimated achievement levels of students in the class. In some instances, instructors base grade cutoffs on a predetermined notion of the knowledge levels required for each grade. In others, grade cutoffs are determined using a "curve," whereby instructors attempt to assign a certain proportion of students to each grade level. Regardless of how grades and the corresponding grade cutoffs are assigned, it is important to emphasize that the manner in which an instructor determines grade cutoffs is not critical in the calculation of the achievement indices. Only the relative ordering of students within classes, as determined by assigned grades, provides information about student achievement.

The following variables are used to model this mechanism for grade generation:

1. The variable y_{ij} denotes the grade assigned to the ith student in the jth class. For notational convenience, the grades used at Duke were coded so that F=1, D–=2, D=3, ..., A+=13. In general, it is assumed that there are C possible grades, ordered from 1 to C.

2. The variable $\mathcal{A}_i$ represents the mean classroom achievement of the ith student, *in classes selected by student i*. This variable is called the *Achievement Index* (AI) of student i. It should be noted that this definition of the "achievement index" differs from the standard definition of a latent trait (e.g., Lord and Novick, 1968) in that the achievement index is defined conditionally for those classes selected by a student.

   The relationship of the achievement index to more standard latent trait models is a subtle one. Perhaps the best way to connect the two concepts is to suppose that a high- or infinite-dimensional trait vector is associated with every student and that a student's expected performance in each class is determined by some weighting of his or her traits. The performance of each student in all classes—whether observed or not—can then be summarized by a weighted trait vector with dimension equal to the number of classes. In standard latent trait analysis, one attempts to estimate every component of the underlying student trait vector from the partially observed vector of weighted classroom traits.

   In contrast, the achievement index represents the mean weighted classroom trait for those classes actually taken by a student. In estimating the achievement index, we are concerned only in estimating the average weighted trait for these classes. We are not interested in estimating the weighted classroom traits for classes not taken, nor are we attempting to estimate the underlying latent traits themselves. Thus, a mathematics major's achievement index represents an entirely different weighting of traits than does an art history major's. By defining the achievement index in this way, the missing data and selection problems which typically confound researchers attempting to measure an actual "personality trait" are circumvented.

3. Grade cutoffs for class j are denoted by $\gamma_0^j, \gamma_1^j, \ldots, \gamma_C^j$. The upper cutoff for an F in class 3 is γ_1^3; for a D–, it is γ_2^3; and so on up to γ_{12}^3, which is the upper cutoff for an A. The upper cutoff for an A+, γ_{13}^j, is ∞, and the lower cutoff for an F, γ_0^j, is $-\infty$.

4. Random variation in the performance of student i in class j is denoted by ϵ_{ij}. This term accounts for the fact that student achievement varies from class to class and that instructor assessment of student achievement is also subject to error. It is assumed that the distribution of each ϵ_{ij} is Gaussian with mean 0 and variance σ_{ij}^2. In addition, the variation of observed student achievement in class j is assumed to depend only on the instructor of class j, denoted by $t(j)$; that is, it is assumed that $\sigma_{ij}^2 = \sigma_{t(j)}^2$, independently of i.

With these variable definitions, the model for grade generation may be summarized as follows. Student i in class j gets a grade of $y_{ij} = k$ if and only if

$$\gamma_{k-1}^{j} < \mathcal{A}_i + \epsilon_{ij} \leq \gamma_k^{j}. \tag{7.3}$$

It follows from (7.3) that the probability that student i receives a grade of k in class j is equal to the area under the normal curve within category k. Letting $\Phi(\cdot)$ denote the cumulative standard normal distribution function (and $\phi(\cdot)$ the standard normal density), this area may be expressed

$$\Phi\left(\frac{\gamma_k^{j} - \mathcal{A}_i}{\sigma_{t(j)}}\right) - \Phi\left(\frac{\gamma_{k-1}^{j} - \mathcal{A}_i}{\sigma_{t(j)}}\right). \tag{7.4}$$

Equation (7.4) describes the likelihood function for a graded response model employing a probit link. By replacing Φ with a logistic distribution function, the graded response model of Young can be obtained; replacing Φ with an arbitrary (but fixed) distribution function defines the class of homogeneous graded response models proposed by Samejima (1972).

The unknown parameters in (7.4) are the vector $\mathcal{A}$ of achievement indices for all students, the vector γ of grade cutoffs for all classes, and the vector σ^2 of error variances for all instructors. If we assume that grades received by students are independent of one another, given model parameters (i.e., local independence), the likelihood function for a cohort of grade data is the product over all probabilities of the form (7.4), or

$$L(\mathcal{A}, \gamma, \sigma^2) = \prod_i \prod_{j \in \mathcal{C}_i} \left[\Phi\left(\frac{\gamma_{y_{ij}}^{j} - \mathcal{A}_i}{\sigma_{t(j)}}\right) - \Phi\left(\frac{\gamma_{y_{ij}-1}^{j} - \mathcal{A}_i}{\sigma_{t(j)}}\right) \right]. \tag{7.5}$$

The first product in (7.5) extends over all students, while the second extends over the set of all classes j taken by student i, denoted here by $\mathcal{C}_i$.

7.2.2   Prior distributions

As in multirater ordinal regression models, the addition of the same constant to all category cutoffs and achievement indices does not affect the likelihood function. Similarly, the likelihood is unchanged if all quantities are divided by the same scalar constant. This is a consequence of the fact that grades are ordinal and possess no natural scale.

One possibility for dealing with this model nonidentifiability is simply to fix the values of two of the category cutoffs: this was the approach taken by Young (1990) in his GRM model. Unfortunately, fixing grade cutoffs can have deleterious effects when the model is used to produce student rankings. To see this, suppose that the lower cutoff for an A+ is fixed at a constant value of, say, 3. Then, the classroom performance of students receiving an A+ in *any* class must be estimated to have a value in excess of 3. But when all grades assigned to students in a given course are A+'s, this assumption results in an inflation of the class rankings of students in that class relative to all other students, even if the average achievement of students taking the class is below average (assuming that average student ability is centered at 0).

A more natural way to establish an underlying scale of measurement is to specify the marginal distribution of the achievement indices, category cutoffs, and error terms. From within the Bayesian paradigm, this is accomplished through the introduction of prior distributions on the unknown quantities of interest.

The prior distributions used here are

$$\mathcal{A}_i \sim N(0, 1), \tag{7.6}$$

$$\epsilon_{ij} \sim N(0, \sigma_{t(j)}^2) \tag{7.7}$$

and

$$\sigma_{t(j)}^2 \sim \text{IG}(\alpha, \lambda) \propto (\sigma_{t(j)}^2)^{-(\alpha+1)} \exp\left(-\frac{\lambda}{\sigma_{t(j)}^2}\right). \tag{7.8}$$

(In classical graded response models, only a prior on the achievement indices is assumed. This prior permits estimation of model parameters to proceed via marginal maximum likelihood, but is not sufficient to make (joint) maximum likelihood estimation feasible.)

Alternative prior densities are considered in Section 7.5. In the baseline model, $\alpha = 1.5$ and $\lambda = 1.5$. Throughout, $N(\mu, \tau^2)$ represents a normal distribution with mean μ and variance τ^2, and IG(α, λ) denotes an inverse-gamma distribution with shape α and scale λ. Recall that the subscript $t(j)$ refers to the teacher of class j.

Note also that the prior density on the achievement indices (7.6) is standard in the psychometric literature and simply serves to fix the scale of measurement. (A similar prior was placed in the latent traits in the IRT model of Chapter 6.)

Equation (7.7) states that the random error associated with the combination of interclass variation in student performance and instructor assessment error follows a normal distribution with mean 0 on the scale of the achievement indices. It is further assumed that the variance of the error depends only on the instructor of class j. This variance is denoted by $\sigma_{t(j)}^2$.

Assumption (7.8) describes the marginal distribution on the variance of the random errors. The particular parameters selected for the inverse-gamma distribution lead to a relatively vague prior on the instructor variances and were chosen to satisfy several criteria. Among these, the positive value of λ eliminates an irregularity in the posterior distribution that occurs when all category cutoffs, achievement indices, and instructor variances assume values close to 0. For $\alpha = \lambda = 1.5$, the priors on each instructor variance have mean 3 and mode 0.6, which seems consistent with the assumption that the marginal distribution on student achievement indices has variance 1.

Defining appropriate prior distributions for the grade cutoffs requires more careful consideration of the mechanisms underlying grade assignment. For example, if a locally uniform prior (subject to the ordering constraint $\gamma_k^j \leq \gamma_{k+1}^j$) is taken on the grade cutoffs, the joint posterior distribution on the achievement indices concentrates its mass above its prior mean of 0. This occurs because unobserved grade categories, which usually correspond to lower grades, are assigned a non-negligible probability in the posterior. In other words, a uniform prior for the grade cutoffs would assign equal an prior probability to all grades, whereas in practice, grades below C are relatively rare. Thus, if a uniform prior were employed for the grade cutoffs, the posterior probability assigned to below C grades, particularly in small classes, would be non-negligible, forcing the posterior distribution on the achievement indices to become highly skewed. As a consequence, the posterior mode estimate would assign zero probability to unobserved grades, while samples from the posterior would assign positive probability to the same grades. The non-negligible mass assigned to low grades would then force the posterior mean of the achievement indices upward toward the inflated grade cutoffs that correspond to the actual grades assigned.

In addition to the effect of unobserved grades on the shape of the posterior, some "standard" noninformative priors on grade cutoffs lead to posterior distributions that are not invariant to shifts in class grades; that is, shifting the grades of all students within a class down by one letter grade (assuming no grades of D– or below) does not change the ordering of students within a class, and thus should not affect the rankings of students. Yet the use of uniform or other vaguely specified priors on the grade cutoffs lead to different estimates of student achievement indices after such shifts.

Priors which lead to posteriors that concentrate negligible mass in unobserved grade categories are therefore needed. Such priors can be specified by introducing binary random variables $(\iota_1^j, \ldots, \iota_K^j) = \iota_j$ to indicate which grade cutoffs correspond to unobserved grades. By placing a high prior probability on the occurrence of unobserved grade categories, the posterior probability assigned to unobserved categories can be made arbitrarily small. One prior density that satisfies these criteria can be specified by first assigning probability 1 to the event $\gamma_{k-1}^j = \gamma_k^j$ whenever $\iota_k^j = 0$, and then taking the prior density on ι_j to be

$$\Pr(\iota_j) = \begin{cases} \varepsilon^{S_j}/[1-(1-\varepsilon)^K] & \text{if } S_j \geq 1 \\ 0 & \text{if } S_j = 0, \end{cases} \qquad S_j = \sum_{k=1}^{K} \iota_k^j, \qquad (7.9)$$

where $\varepsilon \ll 1$.

It follows that the prior conditional distribution for a component of ι_j, given all other components, is

$$\Pr\left(\iota_i^j = 1 \mid \sum_{\substack{k \neq i \\ 1 \leq k < K}} \iota_k^j \geq 1\right) = \frac{\varepsilon}{1+\varepsilon}$$

$$\Pr\left(\iota_i^j = 0 \mid \sum_{\substack{k \neq i \\ 1 \leq k < K}} \iota_k^j = 0\right) = 0. \qquad (7.10)$$

The constraint that $\sum \iota_k^j \geq 1$ ensures that $-\infty = \gamma_0^j \neq \gamma_K^j = \infty$, or in other words, that the lower cutoff for an F cannot equal the upper cutoff for an A+. Of course, whenever a grade of k is observed in class j, the posterior probability that $\iota_k^j = 1$ is 1.

Given ι_j, a noninformative (Jeffreys') prior is assumed for the unique elements of the grade cutoffs in the jth class. Because the achievement indices are assumed to have a $N(0, 1)$ distribution, the Jeffreys' prior on the probability that a student receives a given grade is transformed to this scale, resulting in a prior density of the form

$$p(\gamma^j) \propto \Phi(\gamma_{I_{\min}})^{-1/2} \left\{ \prod_{\substack{i_k \in I \\ i_k > I_{\min}}} [\Phi(\gamma_{i_k}^j) - \Phi(\gamma_{i_{k-1}}^j)] \right\}^{-1/2} \left\{ \prod_{\substack{i_k \in I \\ i_k < I_{\max}}} \phi(\gamma_{i_k}^j) \right\}. \quad (7.11)$$

Here, $I = \{k : \iota_k = 1\}$, i_k denotes the $(k+1)$st largest element of I, and $I_{\min}$ and $I_{\max}$ refer to the smallest and largest elements of I.

7.3 Parameter estimation

In practice, numerical strategies for both sampling from the posterior distribution and for obtaining point estimates of the achievement indices are needed. Sampling strategies are critical for assessing model fit and comparing models, whereas a method for rapidly obtaining point estimates of the achievement indices is important for implementation by, for example, a registrar's office.

7.3.1 Posterior simulation

Sampling from the posterior distribution of model parameters can be accomplished using a Markov chain Monte Carlo algorithm. The particular sampler used in this application is very similar to the hybrid Gibbs sampling/Metropolis-Hastings algorithm for the Bayesian ordinal regression model in Chapter 4.

As in the ordinal regression fitting algorithm, the MCMC algorithm is implemented by introducing variables that represent the unobserved classroom achievement of each student in each class. To this end, define the classroom achievement for the ith student in the jth class to be $Z_{ij} = \mathcal{A}_i + \epsilon_{ij}$. It follows that the likelihood function based on the parameters $(\mathcal{A}, \gamma, \sigma^2)$, augmented with the vector $Z = \{Z_{ij}\}$, may be written

$$L(\mathcal{A}, \gamma, \sigma^2, Z) = \prod_i \prod_{j \in \mathcal{C}_i} \phi\left(\frac{Z_{ij} - \mathcal{A}_i}{\sigma_{t(j)}}\right) I(\gamma_{y_{ij}-1}^j < Z_{ij} \le \gamma_{y_{ij}}^j). \quad (7.12)$$

In (7.12), $I(\cdot)$ represents the indicator function taking a value of 1 if the stated condition is true, and 0 otherwise. Note that integration of (7.12) over the variables $\{Z_{ij}\}$ leads to (7.5). The posterior density of parameters and augmented data is given by

$$g(\mathcal{A}, \gamma, \sigma^2, Z)|\text{data}) \propto L(\mathcal{A}, \gamma, \sigma^2, Z)g(\mathcal{A}, \gamma, \sigma^2), \quad (7.13)$$

where the prior $g(\mathcal{A}, \gamma, \sigma^2)$ is outlined in Section 7.2.2

The posterior distribution that results from these model assumptions can be summarized by simulating from the posterior distribution using the MCMC methodology described in Chapter 5 for the multirater ordinal probit model.

7.3.2   Posterior optimization

Although MCMC algorithms can, in principle, be used to simulate from the posterior distribution described in the previous section, in practice it is not always possible to do so. In the Duke example, there were approximately 10,000 students represented in the 5-year dataset, and these 10,000 students took approximately 6,000 classes during this period. There were also 2,000 instructors, so the total number of model parameters totals roughly 90,000. Clearly, sampling extensively from a model with this many parameters can be computationally very demanding, and even with currently available computer hardware, it is not feasible to do so on a routine basis. From a practical standpoint, it is therefore preferable to find point estimates of the parameters that can be obtained in a relatively short period of time. Such a point estimate is provided by the posterior mode. The posterior mode can be estimated by function maximization, which is generally an order of magnitude faster than integration, at least for smoothly varying functions.

As in the case of related classical GRMs and marginal maximum likelihood, the search for the posterior mode over the parameter space requires judicious choice of initial values. Although a proper prior distribution has been chosen for all parameters, the form of the noninformative priors on the grade-cutoff vectors and the instructor variance parameters results in nonconcavity of the log-posterior density.

Despite this nonconcavity, posterior mode estimates for the student achievement indices $\mathcal{A}$ can be reliably obtained using a variation of the iterated conditional modes (ICM) algorithm, provided that reasonable starting values are used for parameter initialization. In the standard ICM algorithm, the posterior mode is found by sequentially maximizing the conditional distribution of each component of the parameter vector, given current estimates of all other components. This procedure is analogous to a Gibbs sampling algorithm, with the exception that the mode of each conditional posterior distribution is determined at each update, rather than sampling a value from these conditional distributions.

For the Duke data, the ICM algorithm converged within 200 iterations when applied to a single year's data containing 1,500 students, each student taking approximately 35 courses from 2,000 instructors.

7.4   Applications

We now apply the Bayesian graded response model to two datasets. The first is a stylized example borrowed from Larkey and Caulkin (1992) in which the Bayesian achievement indices are shown to produce student rankings that are exactly opposite those obtained using standard GPA. Although this particular example is extreme, the irregularities in its grade assignments are not atypical of actual college transcripts. The example also clarifies the underlying problems associated with GPA-based student assessment and interval-based analyses of ordinal data.

The second example involves a analysis of grades received by a recent class of Duke University undergraduates. The analysis begins with a comparison of AI-derived student class rank and class ranks obtained using unadjusted GPA and additively adjusted GPA. Two performance measures are used in this comparison: one is based on the multiple correlation of each index with external measures of student performance; the second on the predictive success of the indices in ordering student performance within individual classes.

7.4.1   Larkey and Caulkin data

The data in Table 7.1 represent a slight modification of data originally presented by Larkey and Caulkin (1992).

The important feature of this grade data is that all instructors agree that the best ordering of students is I > II > III > IV. Yet, because of differences in instructor grading policies, the observed ranking based on GPA is IV > III > II > I, exactly the *opposite* of the ranking intended by *all* instructors.

To illustrate the properties of the student achievement index, the Bayesian graded response model described in the last section was also applied to this data. Output from the model is displayed in Table 7.2. The columns in this table represent (1) class number, (2) grade received, (3) mean grade assigned in the class, (4) estimated classroom achievement index of the student, (5) mean achievement index of all students enrolled, (6) mean GPA of all students in the class, and (7) estimated category cutoffs for the grade received. Student GPAs appear at the bottom of the columns labeled "Grade," and student achievement indices are listed at the bottom of the column labeled "Estimated Ach'mt." The values of ±∞ were coded as ±9.99. All quantities in the table were normalized so that the posterior mean and variance of the achievement indices were 0 and 1, respectively.

	Student 1	Student 2	Student 3	Student 4	Class GPA
Class 1	B+			B-	3.0
Class 2	C+		C		2.15
Class 3			A	B+	3.65
Class 4	C-	D			1.35
Class 5		A		A-	3.85
Class 6	B+			B	3.15
Class 7		B+	B		3.15
Class 8	B+	B	B-	C+	2.83
Class 9		B	B-		2.85
GPA	2.78	2.86	2.88	3.0	

TABLE 7.1. Larkey and Caulkin example.

Student I. GPA based rank: 4 Achievement Index rank: 1

Course	Grade	Mean Grade	Estimated Ach'mt	Mean Ach'mt	Mean GPA	Grade Cutoffs
CLS 001	B+	3.00	1.25	-0.00	2.89	(-0.00, 9.99)
CLS 002	C+	2.15	1.32	0.36	2.83	(0.24, 9.99)
CLS 004	C-	1.35	1.45	0.79	2.82	(0.59, 9.99)
CLS 006	B+	3.15	1.25	-0.00	2.89	(-0.00, 9.99)
CLS 008	B+	2.83	1.51	0.00	2.88	(0.71, 9.99)
		2.78	2.50	1.15	0.23	2.86

Student II. GPA based rank: 3 Achievement Index rank: 2

Course	Grade	Mean Grade	Estimated Ach'mt	Mean Ach'mt	Mean GPA	Grade Cutoffs
CLS 004	D	1.35	-0.08	0.79	2.82	(-9.99, 0.59)
CLS 005	A	3.85	0.69	-0.36	2.93	(-0.24, 9.99)
CLS 007	B+	3.15	0.80	0.00	2.87	(-0.00, 9.99)
CLS 008	B	2.83	0.36	0.00	2.88	(0.00, 0.71)
CLS 009	B	2.85	0.80	0.00	2.87	(-0.00, 9.99)
		2.86	2.81	0.43	0.09	2.87

Student III. GPA based rank: 2 Achievement Index rank: 3

Course	Grade	Mean Grade	Estimated Ach'mt	Mean Ach'mt	Mean GPA	Grade Cutoffs
CLS 002	C	2.15	-0.69	0.36	2.83	(-9.99, 0.24)
CLS 003	A	3.65	0.08	-0.79	2.94	(-0.59, 9.99)
CLS 007	B	3.15	-0.80	0.00	2.87	(-9.99,-0.00)
CLS 008	B-	2.83	-0.36	0.00	2.88	(-0.71, 0.00)
CLS 009	B-	2.85	-0.80	0.00	2.87	(-9.99,-0.00)
		2.88	2.92	-0.43	-0.09	2.88

Student IV. GPA based rank: 1 Achievement Index rank: 4

Course	Grade	Mean Grade	Estimated Ach'mt	Mean Ach'mt	Mean GPA	Grade Cutoffs
CLS 001	B-	3.00	-1.25	-0.00	2.89	(-9.99,-0.00)
CLS 003	B+	3.65	-1.45	-0.79	2.94	(-9.99,-0.59)
CLS 005	A-	3.85	-1.32	-0.36	2.93	(-9.99,-0.24)
CLS 006	B	3.15	-1.25	-0.00	2.89	(-9.99,-0.00)
CLS 008	C+	2.83	-1.51	0.00	2.88	(-9.99,-0.71)
		3.00	3.30	-1.15	-0.23	2.91

TABLE 7.2. Analysis of Larkey and Caulkin example.

The salient feature of Table 7.2 is that the rank of students based on the achievement index is correct; I is ranked first, II second, III third, and IV fourth.

To gain further insight into the meaning of model parameters, consider the grade of D received by Student II in Class 4. The estimate of this student's achievement in this class was -0.08, which on the probability scale corresponds to the 45th percentile. The mean achievement index for all students taking this class was 0.79, indicating that better than average students were enrolled in the course, and the mean grade assigned in Class 4 was 1.35. Because of the higher-than-average achievement level of students enrolled in Class 4, and the lower-than-average grade assigned, the grade cutoffs for a D were estimated to be between $-\infty$ and 0.59. By comparison, the C cutoffs in Class 2 were $(-\infty, 0.24)$.

7.4.2   A Class of Duke University undergraduates

To illustrate the performance of the Bayesian graded response model for actual college transcripts and to compare its performance to raw GPA and additively adjusted GPA, all models were applied to the grades of a recent class of Duke University undergraduates. Selected transcript summaries from the Bayesian model are displayed in the appendix to this chapter. Transcripts in which the student class rank based on GPA differed sharply from achievement-index-based rank were selectively chosen for display. Approximately 1400 students were ranked. Perusal of these transcripts is left to the reader.

Figure 7.2 is a scatterplot of GPA rank against achievement index rank for this cohort. As illustrated, the correlation between the two measures of student performance is relatively high, and for many students, the differences in class rank obtained from the two indices is not large. However, for other students, the rank percentiles may differ by as much as 40%, and so the effect of this index can be quite substantial.

Given the disparities between GPA rank and achievement-based rank, an obvious question becomes "Which rank better represents student performance?" Two criteria were used to address this question. The first, which appears to be the most commonly used statistic for assessing alternative measures of undergraduate student performance (e.g., Elliot and Strenta, 1988; Young, 1990; Larkey and Caulkin, 1992; and Caulkin, Larkin, and Wei, 1996), was based on the multiple R^2 of the regression of high school GPA, math SAT, and verbal SAT score on each competing measure of student performance. Because the explanatory variables in these regressions are measured independently of college GPA and achievement-based ranks, the resulting R^2 values provide an external measure for model assessment. A potential problem with this measure is that the student selection processes that influence college GPA are also likely to influence high school GPA, and thus might favor models linked to raw GPA adjustments.

For this cohort of undergraduate grades, the multiple R^2 for the regression of college GPA on high school GPA, and math and verbal SAT scores was 0.252. The regression of achievement indices estimated from the ICM algorithm for the same covariates was 0.346, a substantial increase over the value obtained using raw GPA. For the additive adjustment model, the R^2 value was 0.338. This appears to be in general agreement with the value reported in Larkey and Caulkin of 0.321 for a selected subset of Carnegie Mellon undergraduates.

A second criterion for comparing performance indices can be based on the power of the indices in predicting the relative performance of students. For example, an effective index should accurately predict the better of two students, or equivalently, which student is likely to receive the higher grade in a class they take together.

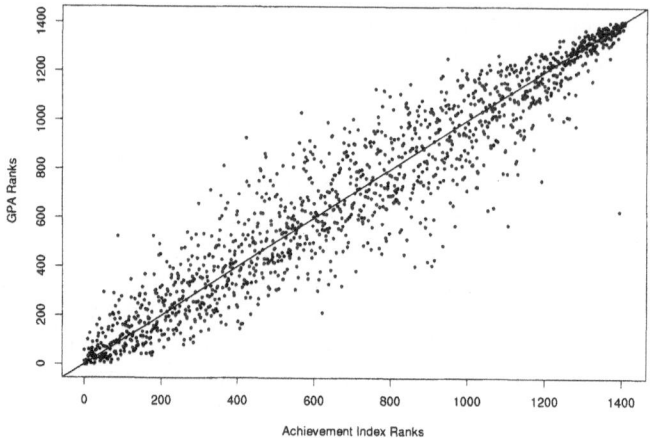

FIGURE 7.2. Scatterplot of student ranks based on GPA versus student ranks based on achievement index for a recent class of Duke undergraduates.

Of course, implementing this criterion requires that the better student be known, and no gold standard exists for making this determination. Complicating the issue further is the fact that there is natural variation in the classroom performance of students, so the better student does not always obtain the higher mark.

Fortunately, the predictive accuracy of achievement indices for pairs of students who receive different grades in each of two courses taken together can be accurately assessed. Supposing that an ideal performance index was available, let the probability that the "better" student receives the higher grade in a randomly selected course be denoted by p. If grades in distinct courses are assumed to be independent, then the probability that the better student, according to the ideal performance index, receives the higher grade in both courses is p^2 and that the better student receives the lower grade in both courses is $(1 - p)^2$. Similarly, the probability that each student gets one of the higher grades is $2p(1 - p)$.

Based on this observation, a simple estimate of p can be obtained by equating $2p(1 - p)$ to the observed proportion of times students in such matched pairs each received one of the higher marks. That proportion was .227 for the Duke cohort, which suggests that an ideal performance index would have an error rate of at least $p = .131$. Thus, .131 provides an estimate of the baseline error rate for *any* index, and we can compare the prediction errors of other indices to this figure, within this subset of student grades. For the Bayesian achievement indices, additively adjusted GPA, and raw GPA, the corresponding error rates were .168, .174, and .206, respectively. Thus, the error rate attributable to model inadequacy was bounded by .037 for the Bayesian achievement indices, while the model-based error for the additively adjusted GPA model was bounded by .043. Raw GPA yielded an model-based error of .075, twice that incurred using the Bayesian achievement indices. As in the case of the R^2 criteria, the Bayesian indices appear to provide the most accurate estimates of student achievement.

7.5   Alternative models and sensitivity analysis

In order to investigate the relative importance of the prior assumptions ((7.6)–(7.11)) in determining final student rankings, several alternative models were examined by varying the prior assumptions and assessing consequent differences in posteriors.

One-component achievement index or two?

Perhaps the most controversial assumption made in the baseline model is that student performance can be adequately summarized through a single achievement index. Note that standard GPA measures and GPA-based class rank are themselves univariate quantities, and so replacing a flawed univariate summary of student performance with a less flawed univariate summary measure seems entirely reasonable. Of course, the value of measuring student performance with any univariate quantity is often questioned, and it is worth examining the loss of information incurred through such summaries.

As a first step toward examining this question, the baseline model was expanded so that the prior distributions on student achievement indices were modeled as bivariate normal random variables having the form

$$\begin{pmatrix} \mathcal{A}_i^1 \\ \mathcal{A}_i^2 \end{pmatrix} \sim N\left(\begin{pmatrix} 0 \\ 0 \end{pmatrix}, \begin{bmatrix} 1 & \rho \\ \rho & 1 \end{bmatrix} \right). \tag{7.14}$$

The value of the prior correlation between components of the achievement indices was assumed to be $\rho = .43$, based on observed correlations between SAT math and verbal scores. Results obtained using a prior correlation of 0 are quite similar to those reported below for $\rho = .43$.

To incorporate the bivariate achievement index into the model for grade generation, each academic department listing in the Duke Registrar's coding of courses was assigned a weight parameter, $w_d, 0 \le w_d \le 1$. In the expanded model, student i is assumed to receive a grade of k in class j in department listing d whenever

$$\gamma_{k-1}^j < w_{d(j)}\, \mathcal{A}_i^1 + (1 - w_{d(j)})\, \mathcal{A}_i^2 \le \gamma_k^j, \tag{7.15}$$

where $w_{d(j)}$ denotes the department offering course j. Beta priors proportional to $w_d(1 - w_d)$ were assumed for each department weight, except in the case of the undergraduate writing course requirement. In order to make the weights and the components of the student achievement indices identifiable, a beta density proportional to $w_d^9(1 - w_d)$ was assigned to the department weight for the undergraduate writing course, which is taken by nearly all incoming Duke students.

Given the department weighting $w_{d(j)}$ for a particular class, the marginal distribution of student achievement indices within a class is no longer $N(0, 1)$, but is, instead, $N(0, \sigma^2_{d(j)})$, where

$$\sigma^2_{d(j)} = w^2_{d(j)} + 2\rho w_{d(j)}(1 - w_{d(j)}) + (1 - w_{d(j)})^2.$$

This change is reflected in an adjustment to the proper prior distribution placed on the grade cutoffs.

The bivariate student achievement indices obtained from a suitably modified ICM procedure were compared to those obtained in the baseline model using the criteria described above. Class ranks were computed for each student from the bivariate index by weighting the two components assigned to each student according to that student's selection of courses. The weighted achievement index for student i was thus defined as

$$A^w_i = \frac{1}{|C_i|} \sum_{j \in C_i} \left(w_{d(j)} A^1_i + (1 - w_{d(j)}) A^2_i \right). \tag{7.16}$$

Class ranks based on $\{A^w_i\}$ were compared against the corresponding ranks obtained from the baseline model. The correlation between the two rankings was found to be quite high (.996), and for all but two or three students, the difference between class rank computed under the two models is relatively small.

The weighted achievement indices were also regressed on high school GPA and verbal and math SAT scores. The multiple R^2 for this regression was .352, somewhat higher than that obtained for the baseline model. In addition, the proportion of orderings correctly predicted for students enrolled in common classes under the two-component model was .148 (using estimated weights and achievement components appropriate for each class), which was just .017 units above the ideal rate of .131. Recall that for the baseline index, the error rate was .168, or 0.037 units above the best obtainable rate.

In terms of accuracy of ranking, the two-component achievement index clearly outperformed the baseline index. However, the gains realized with the two-component index are offset by the additional complexity involved in explaining the two-component index to students, employers, college administrators, and faculty.

Normality of student achievement indices

Another important assumption made within the baseline model is that student achievement indices $\{A_i\}$ are distributed a priori according to a standard normal distribution. In reality, a better prior model for student achievement indices, given that intraclass variation is assumed to follow a normal distribution, would likely involve a mixture distribution on the marginal distribution on achievement indices. However, if such a model were employed, achievement indices would be shrunk toward different values on the achievement scale, depending on the mixture component to which they fell closest. By positing a standard normal distribution on the achievement indices, the prior distribution effectively shrinks all indices toward a common value of 0, which seems more palatable than shrinking student achievement indices toward disparate values. Furthermore, should a more complicated prior model be assumed for the achievement indices, it would likely not have a significant effect on the ranking of students, only the spacing between their estimated indices.

Empirical Bayes estimation of prior densities

The remaining model assumptions concern the particular inverse-gamma distribution employed as the prior density on the instructor variances and the form of the prior density specified for the grade cutoffs. To investigate the impact of these assumptions on final student ranking, alternative estimates of achievement indices were obtained using empirical Bayes methodology and more standard "uniform" priors.

In empirical Bayes methodology, a prior distribution is specified with a set of unknown hyperparameters. Instead of expressing the uncertainty in the hyperparameters using a prior in a fully Bayesian approach, the empirical Bayes approach estimates these hyperparameters from the observed data. This approach is attractive because the values of unknown parameters are empirically estimated from data and are thus, in some sense, more "objective" than the usual prior specification. This approach also circumvents the potentially difficult task of specifying a prior distribution.

From an empirical Bayes viewpoint, a prior density on the distribution of grade cutoffs can be obtained by letting π_j denote the probability vector describing the multinomial probabilities that students in class j are assigned to different grade categories and assuming that π_j is drawn from a Dirichlet distribution with parameters $\omega = (\omega_1, \ldots, \omega_K)$; that is, assume that

$$p(\pi_j) \propto \prod_{k=1}^{C} (\pi_{j.k})^{\omega_k - 1}, \tag{7.17}$$

independently in j. Equation (7.17) provides an achievement-index-free model for each class's multinomial probability vector describing grade assignments. Using the observed number of grades assigned in all classes, it is straightforward to obtain the maximum likelihood estimate for ω, denoted here by $\hat{\omega}$. By substituting this estimate in (7.17) and transforming to the standard normal scale, one can obtain an "empirical Bayes" prior on the vector of grade cutoffs.

With the use of this prior and the standard normal prior on the achievement indices, it is possible to generate sample transcripts of students for any chosen value of instructor variance parameters (α, λ). By simulating student transcripts, an empirical Bayes method-of-moments approach can also be used for estimating values of the instructor variance parameters. More specifically, both the sample mean of the within-student variance of assigned grades and the variance of this mean student variance between students can be calculated directly from sampled student transcripts. By appropriate choice of α and λ, sample transcripts can be generated to mimic the observed values of these quantities. Implementing this moment-matching procedure yielded approximate values for α and λ of 6.0 and 5.7, respectively.

The sample correlation between the ICM estimates of student ranks based on achievement indices estimated from the model employing empirical Bayes priors and the baseline model was quite high—0.986. However, for a small minority of students, the differences in class rank were as great as 20%, which again raises the question of which model should be preferred. For the empirical Bayes estimates, the multiple R^2 of the regression of the indices on SAT and high school GPA was 0.302, and the error in predicting the grades of paired students in common classes was 0.170. Both measures suggest that the baseline model was the more accurate, and further evidence of this assertion was obtained through residual analyses of the type performed for the baseline model.

The comparatively poor performance of the empirical Bayes estimates of student achievement is somewhat surprising, although a partial explanation for this failure may be found by examining the relationship between grading policies of instructors and classroom attributes. For example, an association was found between class size and mean grade, and this suggests that the failure of the empirical Bayes model might be caused in part by its failure to account for decreasing trends in mean grade with class size. It is likely that similar trends exist with other course attributes. Apparently, partially misspecified empirical Bayes priors are less effective in modeling grade generation than the relatively vague prior employed in the baseline model. These observations support the hypothesis that interval grade information cannot easily be incorporated into student assessments.

Locally uniform priors

Locally uniform priors can be stipulated for both the grade cutoffs and the instructor-variance hyperparameters α and λ. Unfortunately, independent uniform priors on α and λ cause the posterior distribution to take its maximum value near the origin, making ICM estimation futile.

As an alternative to full ICM estimation for such models, one may, instead, fix the values of the variance hyperparameters near their posterior mean and maximize the remaining model parameters as before. The posterior mean of (α, λ) can be approximated in a straightforward way by modifying the Bayesian data augmentation-MCMC algorithm so that values of these parameters are sampled along with all others. Such a procedure is analogous to marginal maximum likelihood procedures, although in this case, "marginalization" is performed on the hyperparameters (α, λ) instead of the achievement indices (i.e., latent examinee abilities).

Two models employing locally uniform priors were studied. In the first, denoted U1, the grade cutoffs were assumed to have the prior distributions specified in (7.8), but independent uniform priors were assumed for both α and λ. In the second model, U2, uniform priors on both the grade cutoffs (subject to the ordering constraint $\gamma_k^j \leq \gamma_{k+1}^j$), and the hyperparameters α and λ were employed. For numerical stability, grade cutoffs were assumed to be uniformly distributed within the interval $(-10, 10)$.

The posterior mean of (α, λ) under these two models were $(3.8, 4.6)$ (U1) and $(9.7, 2.8)$ (U2).

Interestingly, the inverse-gamma prior employed in model U2 led to a posterior that concentrated its mass closer to 0 than did the other models considered. This was due to the interaction between the priors on the achievement indices and the grade cutoffs. Because of the non-negligible posterior probability assigned to the unobserved grade categories under model U2, A and B categories were forced to take cutoffs well above 0, whereas the $N(0,1)$ prior on the achievement indices pulled the achievement indices toward 0. As a consequence, the posterior mean of achievement indices was approximately 0.8, and the posterior variance were 0.2. Instructor variance parameters decreased accordingly. In contrast, the posterior mean and variance of the achievement indices under model U1 were 0.2 and 1.2, respectively.

The multiple R^2 statistics for U1 and U2 were .342 and .351; the prediction errors for paired students and classes were .168 and .171. The correlations of the student ranks obtained from models U1 and U2 with the baseline model's ranks exceeded 0.999 and 0.998, respectively.

The high correlation of ranks obtained under the various models of this section indicate that estimates of student achievement are robust to minor variations in model assumptions. Indeed, differences in student ranks obtained using these differing model assumptions is small compared to the posterior uncertainty of the same ranks obtained using any particular model. This point is illustrated in Figure 7.3, which depicts the posterior standard deviation of each student rank from the Duke cohort in the baseline model versus the absolute difference of ranks that were obtained using the baseline model and model U2. This figure demonstrates that the posterior standard deviation associated with each student's rank is larger, and in most cases much larger, than the difference attributable to model specification. In fact, on average, the two ranks obtained from the baseline model and model U2 were less than one-fifth of a standard deviation apart. This shows that the posterior variance associated with achievement indices dwarfs the uncertainty associated with the specific choice of prior hyperparameters and that the relatively vague priors chosen in the baseline model likely provide nearly optimal estimates of class rank among one-component models.

7.6   Discussion

The application of graded response model methodology to student grades illustrates the great potential that ordinal data models have to improve measurements of quantities important to society at large. By more accurately measuring such quantities, better incentives for human performance can be introduced and better products can be brought to market.

From a technical standpoint, the primary innovations of the Bayesian model over previously proposed GRM models are the introduction of prior distributions on instructor variances and grade cutoffs, and the construction of a prior distribution on the grade cutoffs. The priors on the instructor variance parameters and grade cutoffs replace the hard constraints on grade cutoffs employed in earlier models and allow model parameters to more accurately adjust to observed grading pat-

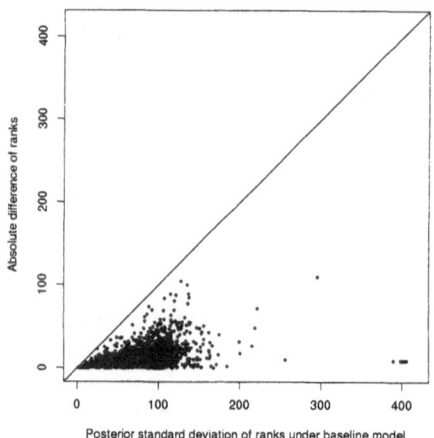

FIGURE 7.3. Posterior variability of AI-based class rank. This figure depicts the posterior standard deviation of the rank of each Duke student under the baseline model versus the absolute difference of the ICM ranks obtained under the baseline model and model U2. The area under the line contains those points at which the differences in rank were smaller than the posterior standard deviation of the rank under the baseline model. No points fell above the line.

terns. The proposed model for grades is essentially able to discard all extraneous interval information contained in observed grade data and, instead, relies on only the rankings of students within classes. As a result, shifting all grades assigned to a class of students up or down has no effect on estimated student achievement indices. An instructor who assigns A's to every student in a class is no longer doing the students in his or her class a favor; such a grading policy has no effect on the achievement indices of any of the students in the class. Likewise, instructors who assign grades that are significantly below the university average no longer penalize students in their classes.

From a computational standpoint, the inclusion of proper priors on the rater variance (i.e., item discrimination) parameters is important because it makes posterior optimization and MAP estimation a reasonable strategy for finding point estimates of model parameters. Without the inverse-gamma priors assumed for the rater variances, the MAP estimate collapses to zero.

7.7   Appendix: selected transcripts of Duke University undergraduates

The columns in the transcripts below represent (1) course designator, (2) grade, (3) mean grade in class, (4) estimated classroom achievement index of the student,

(5) mean achievement index of all students enrolled in the class, (6) mean GPA of students in the class, and (7) estimated category cutoffs for the grade received. Approximately 1400 students were ranked.

Student 1. GPA based rank: 296 Achievement Index rank: 822

Course	Grade	Mean Grade	Estimated Ach'mt	Mean Ach'mt	Mean GPA	Grade Cutoffs
CHM 01--	B-	2.90	-0.42	0.02	3.23	(-0.61,-0.25)
MUS 05--	B+	3.74	-1.46	0.00	3.46	(-9.99,-0.89)
PHL 04--	B	2.88	-0.58	-0.36	3.26	(-0.86,-0.31)
UWC 00--	B+	3.72	-2.01	0.04	3.30	(-9.99,-1.25)
BIO 04--	A	4.00	-0.23	0.33	3.43	(-9.99, 9.99)
CHM 01--	B	2.95	0.00	0.08	3.25	(-0.26, 0.30)
ECO 05--	B+	2.86	0.36	0.16	3.35	(0.16, 0.57)
MTH 03--	B-	2.67	-0.27	-0.02	3.23	(-0.35,-0.20)
ECO 14--	B+	3.30	-0.23	-0.23	3.59	(-9.99, 9.99)
GER 01--	A	3.85	1.11	-0.13	3.29	(-0.16, 9.99)
STA 11--	A-	3.28	0.68	0.05	3.33	(0.43, 1.00)
CPS 01--	A	3.71	-0.05	-0.29	3.22	(-0.59, 0.56)
CST 14--	A	3.04	1.65	-0.18	3.24	(0.95, 9.99)
ECO 14--	B	3.47	-1.21	0.22	3.45	(-9.99,-0.35)
GER 06--	A-	3.80	-0.36	0.51	3.49	(-9.99, 0.94)
PE 10--	A	4.00	-0.62	-0.34	3.14	(-9.99, 1.05)
ECO 06--	A-	3.70	-0.23	-0.23	3.59	(-9.99, 9.99)
GER 11--	A	4.00	-0.23	-0.23	3.59	(-9.99, 9.99)
GER 11--	A-	3.70	-0.23	-0.23	3.59	(-9.99, 9.99)
PS 10--	A-	3.70	-0.23	-0.23	3.59	(-9.99, 9.99)
GER 15--	A	4.00	-0.23	-0.23	3.59	(-9.99, 9.99)
GER 15--	A	4.00	-0.23	-0.23	3.59	(-9.99, 9.99)
HST 10--	A	4.00	-0.23	-0.23	3.59	(-9.99, 9.99)
PS 10--	A	4.00	-0.23	-0.23	3.59	(-9.99, 9.99)
	3.59	3.55	-0.23	-0.08	3.41	

Student 2. GPA based rank: 822 Achievement Index rank: 350

Course	Grade	Mean Grade	Estimated Ach'mt	Mean Ach'mt	Mean GPA	Grade Cutoffs
CHM 01--	A+	2.90	2.54	0.02	3.23	(2.38, 9.99)
LAT 06--	A	3.65	2.01	0.65	3.46	(0.42, 9.99)
MTH 10--	A	2.76	2.01	0.79	3.48	(1.59, 9.99)
UWC 00--	A	3.27	1.66	-0.69	2.97	(0.97, 9.99)
BIO 12--	B+	3.10	0.48	0.63	3.44	(-0.15, 1.08)
CHM 01--	A	2.95	1.44	0.08	3.25	(1.11, 2.56)
LAT 06--	B+	3.57	-1.36	0.17	3.22	(-9.99,-0.47)
MTH 10--	C+	3.14	0.04	1.05	3.61	(-0.34, 0.32)
CHM 15--	A	2.83	0.86	0.09	3.12	(0.09, 9.99)
CA 14--	A-	3.70	0.62	0.62	3.21	(-9.99, 9.99)
LAT 10--	B	3.00	0.16	0.07	3.12	(-0.37, 0.61)
PHY 05--	B-	2.85	0.17	0.37	3.35	(0.05, 0.28)
STA 11--	D	2.47	-1.16	0.24	3.25	(-9.99,-0.74)
BIO 16--	D+	2.98	-1.67	0.48	3.36	(-9.99,-1.44)
CA 14--	C+	2.30	0.62	0.62	3.21	(-9.99, 9.99)
LAT 10--	D-	1.00	0.62	0.62	3.21	(-9.99, 9.99)
PHY 05--	D	2.56	-1.09	0.51	3.39	(-9.99,-0.85)
ARB 00--	A	4.00	0.62	0.56	3.42	(-9.99, 9.99)
BIO 15--	B-	2.86	0.28	0.41	3.32	(0.03, 0.51)
BIO 18--	A	2.62	1.85	0.36	3.24	(1.47, 9.99)
SP 00--	B	3.00	0.62	0.62	3.21	(-9.99, 9.99)
ARB 00--	A	4.00	0.62	0.62	3.21	(-9.99, 9.99)
BIO 19--	A	4.00	0.62	0.62	3.21	(-9.99, 9.99)
REL 15--	A+	3.74	1.65	-0.09	3.25	(1.00, 9.99)
SP 00--	A-	3.37	1.14	0.35	3.36	(-0.32, 9.99)
BIO 15--	A	3.89	1.45	0.44	3.38	(-0.04, 9.99)
BIO 27--	A-	3.70	0.62	0.62	3.21	(-9.99, 9.99)
ARB 06--	A-	3.70	0.62	0.62	3.21	(-9.99, 9.99)
ARB 19--	A	4.00	0.62	0.62	3.21	(-9.99, 9.99)
BIO 22--	A	3.43	1.12	0.46	3.41	(0.67, 9.99)
SP 06--	B+	3.80	-1.44	0.10	3.33	(-9.99,-0.79)
ARB 19--	B	3.50	-0.57	0.64	3.35	(-9.99, 0.50)
BIO 19--	C-	1.70	0.62	0.62	3.21	(-9.99, 9.99)
SP 07--	C-	2.20	-0.50	0.18	3.07	(-9.99, 0.15)
SP 13--	A	4.00	0.62	0.62	3.21	(-9.99, 9.99)
SP 13--	A	4.00	0.62	0.62	3.21	(-9.99, 9.99)
		3.21	3.18	0.62	0.42	3.27

Appendix: Software for Ordinal Data Modeling

An Introduction to MATLAB

MATLAB is an interactive high-level language for integrating computation, visualization, and programming. It was originally written to provide an easy access to matrix software developed by the LINPACK and EISPACK projects. The basic data element in MATLAB is a matrix. One performs calculations by entering calculator-type instructions in the Command Window. Alternatively, one can execute sets of commands by means of scripts called M-files and functions. MATLAB has extensive facilities for displaying vectors and matrices as graphs.

A toolbox of MATLAB version 5 functions written by the authors is available at the web site http://www-math.bgsu.edu/~albert/ord_book/ This set of functions can be used together with the MATLAB software (available from The MathWorks, Inc., 24 Prime Park Way, Natick, MA 01760-1500) to perform all of the calculations and graphics that are illustrated in this book. Here, we outline the use of some of the MATLAB functions in this toolbox for fitting and criticizing binary, ordinal, multirater, and item response regression models.

Chapter 2 — Review of Bayesian computation

Defining the posterior density

The MATLAB functions listed below can be used to implement the different summarization methods for a posterior density with two parameters. To use these functions, a MATLAB function needs to be written which defines the logarithm of the posterior density. The following function logpost2 computes the log of the posterior density for the log odds ratio and log odds product discussed in Section 2.2:

```
function val=logpost2(xy,data)
t1=xy(:,1);    t2=xy(:,2);
y1=data(1); n1=data(2); y2=data(3); n2=data(4);
g1=(t1+t2)/2; g2=(t2-t1)/2;
val=y1*g1-n1*log(1+exp(g1))+y2*g2-n2*log(1+exp(g2));
```

There are two inputs to this function. The matrix xy contains values of the parameter vector, where each row corresponds to a set of values for the two parameters, and the vector data contains values of data and prior parameters that are used in the definition of the posterior density. The output val is a vector of values of the logarithm of the density at the parameter values specified by xy.

The multivariate normal approximation

```
[mode,var]=laplace('logpost2',mode,numiter,data)
```

The function laplace implements the multivariate normal approximation described in Section 2.3. There are four inputs: 'logpost2' is the name of the function which defines the log posterior, mode is a guess at the posterior mode, numiter is the number of iterations of the algorithm for finding the posterior mode, and data is a vector of prior parameters and data that is used by the function 'logpost2'. The output is the posterior mode mode and the approximate variance-covariance matrix var.

Grid integration

```
[lint,mom]=ad_quad2('logpost2',mom,numiter,data)
```

The function ad_quad2 implements the adaptive quadrature algorithm described in Section 2.3. One inputs the name of the function 'logpost2' which defines the log posterior, a vector mom of guesses at the posterior moments of the distribution, the number of iterations numiter of the algorithm, and the vector data of data values used by 'logpost2'. The output of the function is the logarithm of the integrand of the density lint and a vector mom of estimates at the posterior moments of the density.

Gibbs sampling

```
sim_sample=gibbs('logpost2',start,m,scale,data)
```

The function gibbs illustrates the Gibbs sampling algorithm, where a Metropolis-Hastings algorithm is used to sample from each full-conditional distribution. The inputs are the definition 'logpost2' of the log posterior density, a vector `start` which contains the starting value of the algorithm, the number of iterations m of the algorithm, a vector `scale` of scale constants used in the MH algorithms, and the vector `data` of data values used in `'logpost2'`. The output is a matrix of simulated values from the posterior density, where each row corresponds to a different simulated value.

Chapter 3 — Regression models for binary data

Data setup

Data for a regression problem with a binary response are stored as a matrix. Each row of the matrix corresponds to a single experimental unit. The first column contains the proportion of successes, the second column contains the corresponding sample size, and the remaining columns of the matrix contain the set of regression covariates. For the statistics class dataset described in Chapter 3, each experimental unit corresponds to a binary response, and the MATLAB matrix `stat` which stores this data are given as follows:

```
stat=[
     0    1    1    525
     0    1    1    533
     1    1    1    545
     0    1    1    582
     1    1    1    581
     1    1    1    576
     1    1    1    572
     1    1    1    609
     1    1    1    559
     1    1    1    543
     1    1    1    576
     1    1    1    525
     1    1    1    574
     1    1    1    582
     1    1    1    574
     0    1    1    471
     1    1    1    595
     0    1    1    557
     0    1    1    557
     1    1    1    584
     1    1    1    599
     0    1    1    517
     1    1    1    649
     1    1    1    584
     0    1    1    463
     1    1    1    591
     0    1    1    488
     1    1    1    563
     1    1    1    553
     1    1    1    549];
```

Maximum likelihood estimation and model criticism

```
[beta,var,fitted_probs,dev_df,pearson_res,dev_res,adev_res]
    = breg_mle(stat,'l');
```

The MATLAB function breg_mle finds the maximum likelihood estimate of the vector of regression coefficients. The input to this function is the data matrix stat and the link function ('l' for logit, 'p' for probit, and 'c' for complementary log-log). As output, the vector beta contains the mle and the matrix var contains the associated variance-covariance matrix of the estimate. In addition, the function outputs the vector of fitted probabilities fitted_probs, a vector dev_df containing the deviance and associated degrees of freedom, the vector of Pearson residuals pearson_df, the vector of deviance residuals dev_res, and the vector of adjusted deviance residuals adev_res.

Bayesian fitting using a flat prior

```
[Mb,arate]=breg_bay(stat,m,'l');
```

The MATLAB function breg_bay takes a simulated sample from the posterior distribution of the regression coefficients (under a flat prior) using the Metropolis-Hastings algorithm. There are three inputs to this function: stat is the data matrix, m is the number of iterations of the simulation algorithm, and 'l' indicates that a logistic link will be used. The output is the matrix Mb, where each row of the matrix corresponds to a simulated value from the posterior distribution, and the constant arate which is equal to the acceptance rate of the Metropolis algorithm.

Posterior fitted probabilities and residual distributions

Several functions are available for summarizing the simulated sample from the posterior distribution.

```
[fit_prob,residuals]=lfitted(Mb,stat,'l');
```

The function lfitted computes summaries of the posterior distributions of the fitted probabilities $\{\hat{p}_i\}$ and the residuals $\{y_i - \hat{p}_i\}$. One inputs the matrix of simulated values Mb, the data matrix stat, and the link function 'l'. One output is the matrix fit_prob, where each row of the matrix corresponds to the 5th, 50th, and 95th percentiles of the posterior distribution of p_i. Similarly, a row of the output matrix residuals contains the 5th, 50th, and 95th percentiles of the posterior distribution of the Bayesian residual $y_i - p_i$.

```
plotfitted(stat(:,4),fit_prob,'SAT', 'Probability')
```

The function plotfitted is useful for graphing the summaries of the posterior distribution that are produced by the function lfitted as shown in Figure 3.5. The inputs to this function are the vector stat(:,4) that will be plotted on the horizontal axis, the summary matrix fit_prob of the posterior distribution of the fitted probabilities, and the strings `'SAT'` and `'Probability'` which are used to label the horizontal and vertical axes on the plot.

Computation of latent residuals

```
[log_scores,sz]=llatent(Mb,stat);
```

The function `llatent` computes the posterior means of the ordered latent residuals (assuming a logistic link) discussed in Section 3.4. This function also produces the logistic scores plot as shown in Figure 3.9. The input to this function is the matrix of simulated values from the posterior `Mb` and the data matrix `stat`. The output is the vector of logistic scores `log_scores` and the vector of posterior means of the ordered latent residuals `sz`.

Bayesian probit fitting with a flat prior using data augmentation

```
Mb=b_probg(stat,m);
```

The function `b_probg` implements the Gibbs sampling algorithm for fitting a Bayesian binary regression model with a probit link. There are two inputs to this function: `stat`, the data matrix, and `m`, the number of iterations of the simulation. The output matrix `Mb` contains the simulated sample from the posterior distribution, where the rows of the matrix correspond to the simulated variates.

Bayesian fitting using an informative prior

```
[Mb,arate]=breg_bay(stat,m,'l',prior);
```

The function `breg_bay` can also be used to simulate from the posterior distribution using a conditional means informative prior distribution. There are four inputs to this function: the data matrix `stat`, the number of iterations of the algorithm `m`, the string `'l'` indicating a logistic link, and a matrix `prior` which contains the parameters of the conditional means prior. The output of this function is the matrix `Mb` of simulated values from the posterior and the acceptance rate `arate`.

In the example of Section 3.2.4, the prior information was that a student with covariate vector [1 500] would pass with probability .3, a student with covariate vector [1 600] would pass with probability .7, and each statement was worth five observations. This prior information is inputted by means of the matrix

```
prior=[.3 5 1 500
       .7 5 1 600];
```

Computation of a marginal likelihood

```
[beta,var,lmarg]=cmp(stat,prior);
```

The function cmp computes the marginal likelihood used in computing the Bayes factor discussed in Section 3.4. The input of this function is the data matrix stat and a matrix prior, which contains the parameters of a conditional means prior distribution. (The specification of the matrix prior is illustrated above.) The function has three outputs: the posterior mode beta, the associated posterior variance-covariance matrix var, and the natural logarithm of the value of the marginal likelihood lmarg.

Computation of a posterior predictive distribution

```
p_std=post_pred(Mb,conduct);
```

The function post_pred obtains a simulated sample from the posterior predictive distribution of the standard deviation of the counts $\{y_i^*\}$ as discussed in Section 3.5. The input to this function is the matrix of simulated values Mb from the posterior distribution of the regression vector and the dataset matrix conduct. The output vector p_std represents a simulated sample from the posterior predictive distribution of the standard deviation of $\{y_1^*, ..., y_n^*\}$.

Fit of a random effects model

```
[Mbeta,Ms2,p_std2]=logit_re2(conduct,m,ab);
```

The function logit_re2 fits the Bayesian random effects model discussed in Section 3.5. The inputs are the dataset conduct, the size of the simulated sample m, and the vector ab of parameter values for the inverse-gamma distribution placed on the random effects variance. The function returns the matrix Mbeta of simulated values from the marginal posterior distribution of the regression vector β, the vector Ms2 of simulated values from the marginal posterior distribution of the random effects variance σ^2, and the vector p_std of simulated values from the posterior-predictive distribution of the standard deviation of $\{y_1^*, ..., y_n^*\}$.

Chapter 4 — Regression models for ordinal data

Data setup

The data input for a ordinal response regression problem are a single matrix. Each row of the matrix corresponds to a single experimental unit. The first column contains the ordinal responses and the remaining columns contain the set of regression covariates. For the statistics class dataset described in Chapter 4, the MATLAB matrix ostat which stores this data are given as follows:

```
ostat=[
    2    1    525
    2    1    533
```

```
4   1   545
2   1   582
3   1   581
...
4   1   563
4   1   553
5   1   549]
```

Maximum likelihood estimation

```
[mle,cov,dev,devRes,fits] = ordinalMLE(ostat,C,link);
```

The function `ordinalMLE` finds the maximum likelihood estimate of the unknown category cutoffs and the regression coefficients. The inputs to this function are the data matrix `ostat`, the number of categories C, and a string variable `link` indicating the link function. The function outputs the maximum likelihood estimate `mle`, the associated variance-covariance matrix `cov`, the deviance statistic `dev`, the vector `devRes` containing the signed deviance contributions, and a matrix of fitted probabilities `fits`.

Bayesian fitting using flat or informative priors

```
[sampleBeta, meanZ, accept]
            = sampleOrdProb(ostat,C,mle,m,prior);
```

The function `sampleOrdProb` takes a simulated sample from the posterior distribution of the regression coefficients and category cutoffs using the MCMC algorithm described in Section 4.3.2. The inputs to this function are the data matrix `ostat`, the number of categories C, the vector `mle` containing the maximum likelihood estimates (found using the function `ordinalMLE`), and the number of simulated values `m`. The default is a flat prior for the unknown parameters. If an informative prior is used, `prior` is the name of a function that returns a value of the prior density at the parameter vector. The output is a matrix `sampleBeta` of simulated values where the rows correspond to the m samples and the columns correspond to the components of `mle`. In addition, the vector `meanZ` contains the means of the sample latent variables by observation and `accept` contains the MH acceptance rate for the category cutoffs.

Chapter 5 — Analyzing data from multiple raters

Data setup

The data input for analyzing multiple rater data consist of two matrices. The matrix N contains the observed ordinal responses, where the rows correspond to the "items" and the columns to the "raters." When applicable, the matrix X contains the design matrix for the regression model, where the rows correspond to the items and the columns to the regression parameters.

Bayesian fitting of a multiple rater model

```
[Zij, Z, S, Cats, accept]
        = sampleMulti(N,sampSize,alpha,lambda);
```

The function sampleMulti obtains a simulated sample from the posterior distribution of the multirater model described in Section 5.2. The inputs are the data matrix N, the number of iterates of the simulation sampSize, and the parameters alpha and lambda of the inverse-gamma distribution placed on the rater variances. The output is the matrix of simulated values Zij of the rater-item values $\{t_{ij}\}$, the matrix Z of simulated values of the latent traits Z, the matrix S of simulated variates from the rater variances, the matrix Cats of simulated values from the category cutoffs γ, and the acceptance rate accept of the MH algorithm used to sample the category cutoffs.

Bayesian fitting of a multiple rater model (regression case)

```
[Zij,Z,S,Cats,B,Sr,accept]
        = sampReg(N,X,bZ,sampSize,alpha,lambda);
```

The function sampReg obtains a simulated sample from the posterior distribution of the multirater model with regression described in Section 5.3. The inputs to this function include the data matrix N, the number of iterates sampSize, and the hyperparameters alpha and lambda as described for the previous function. In addition, this function requires the input of the design matrix X and the vector bZ, which contains the estimates of the mean latent traits. The output is Zij, Z, S, Cats, and accept, as described earlier; also, the function outputs B, a matrix containing simulated values of the posterior of the regression parameter, and Sr, a vector containing simulated values of the posterior of the regression variance.

Bayesian fitting of a ROC analysis

```
[Zij,Z,S,Cats,m0,m1,v0,v1]
        = roc(N,D,TmT,sampSize,alpha,lambda);
```

The function roc obtains a simulated sample from the posterior distribution for the ROC analysis described in Section 5.4. The input to this function is the data matrix N, the vector D, which indicates the patients' disease status (1–yes, 0–no), the vector TmT which indicates the treatments for all patients (0's or 1's), the number of MCMC iterates sampSize, and the prior parameters on the rater parameters alpha and lambda. The output is the matrix Zij containing the simulated rater-item values from their posterior distribution, the matrix Z of simulated values of the latent item traits, the matrix S of simulated values from the posterior on the rater variances, and the matrix Cats of sampled values from the category cutoffs. Also, m0 and m1 are vectors containing sampled values of the disease means, v0 and v1 are vectors of the sampled disease variances, and accept is the acceptance probability for the sampling algorithm on the category cutoffs.

Chapter 6 - Item response modeling

Data setup

The data are stored in the matrix form described in Section 6.3.1. The matrix has n rows and k columns, where the rows correspond to the individuals that are judged and the columns correspond to the judges or items. The entries of the matrix are 0's and 1's, where (in the exam example) 0 indicates an incorrect response and 1 indicates a correct response. In the following, the matrix ratings contains the ratings of the student judges on the shyness of their peers.

Bayesian fit of two-parameter probit item response model

```
[av,bv,th_m,th_s] = item_r(ratings,m_a,s_a,s_b,m);
```

The function item_r takes a simulated sample of the posterior distribution of the item and ability parameters of the two-parameter item response model with probit link described in Section 6.5. The input to the function is the data matrix ratings, the mean m_a and standard deviation s_a of the item slope parameters, the standard deviation s_b of the item intercept parameters, and the number of iterations m of the Gibbs sampling algorithm. The function returns the matrix av of simulated values from the marginal posterior distribution of the item slope parameters, where each row of the matrix corresponds to a single sampled vector. In addition, bv is the matrix of simulated values of the item intercept parameters, th_m is a vector containing the posterior means of the ability parameters, and th_s contains the respective posterior standard deviations of the ability parameters.

Posterior estimates of the item response curve

```
[pr,lo,hi] = irtpost(av,bv,theta);
```

The function irtpost computes summaries of the posterior densities of all of the item response curves. The inputs to the function are the matrices av and bv of simulated values from the posterior distributions of the item slope and intercept parameters, and a vector theta of values of the ability parameter. The function outputs a matrix pr of posterior medians of the probability of correct response, where the rows of the matrix correspond to the values of the ability parameter and the columns correspond to the different items. Also, the matrices lo and hi contain the respective 5th and 95th percentiles of the posterior distribution of the probability of a correct response.

Bayesian fit of one-parameter probit item response model

```
[bv,th_m,th_s] = item_r1(ratings,s_b,m);
```

The function item_r1 takes a simulated sample of the posterior distribution of the item and ability parameters of the one-parameter item response model with probit link described in Section 6.8. The input to the function is the data matrix ratings, the standard deviations_b of the item intercept parameters, and the number of iterations m of the Gibbs sampling algorithm. The function returns the matrix bv of simulated values of the item intercept parameters, the vector th_m, containing the posterior means of the ability parameters, and the vector th_s, containing the respective posterior standard deviations of the ability parameters.

Bayesian fit of two-parameter probit item response model with an exchangeable prior

```
[av,bv,th_m,th_s,av_m,av_s2] = item_r_h(ratings,ab,m);
```

The function item_r_h takes a simulated sample of the posterior distribution of the item and ability parameters of the two-parameter item response model with an exchangeable prior discussed in Section 6.10. The input to the function is the data matrix ratings, the vector ab of parameters of the inverse-gamma prior for the variance τ^2, and the number of iterations m of the Gibbs sampling algorithm. The function returns the matrix av of simulated values from the marginal posterior distribution of the item slope parameters, the matrix bv of simulated values of the item intercept parameters, the vector th_m, containing the posterior means of the ability parameters, and the vector th_s, containing the respective posterior standard deviations of the ability parameters.

References

Abramowitz, M. and Stegun, I.A., editors (1972), *Handbook of Mathematical Functions*, New York: Dover Publications Inc.

Albert, J.H. (1992), "Bayesian Estimation of Normal Ogive Item Response Curves Using Gibbs Sampling," *Journal of Educational Statistics*, **17**, 261-269.

Albert, J.H. (1996), *Bayesian Computation Using Minitab*, Belmont, CA: Duxbury Press.

Albert, J.H. and Chib, S. (1993), "Bayesian Analysis of Binary and Polychotomous Response Data," *Journal of the American Statistical Association*, **88**, 669-679.

Albert, J.H. and Chib, S. (1995), "Bayesian Residual Analysis for Binary Response Regression Models," *Biometrika*, **82**, 747-759.

Antleman, G. (1997), *Elementary Bayesian Statistics* Cheltenham, Hong Kong: Edward Elgar Publishing.

Baker, F. B. (1992), *Item Response Theory: Parameter Estimation Techniques*, New York: Marcel Dekker.

Bedrick, E.J., Christensen, R., and Johnson, W. (1996), "A New Perspective on Priors for Generalized Linear Models," *Journal of the American Statistical Association*, **91**, 1450-1460.

Bernardo, J.M. and Smith, A.F.M. (1994), *Bayesian Theory*, New York: John Wiley & Sons.

Berry, D. (1995), *Statistics: A Bayesian Perspective*, Belmont, CA: Duxbury Press.

Bock, R.D. and Aitken, M. (1981), "Marginal Maximum Likelihood Estimation of Item Parameters: Application of An EM Algorithm," *Psychometrika*, **46**, 443-459.

Bock, R.D. and Lieberman, M. (1970), "Fitting a Response Model for *n* Dichotomously Scored Items," *Psychometrika*, **35**, 179-197.

Box, G.E.P. and Tiao, G.C (1973), *Bayesian Inference in Statistical Analysis*, Reading, MA: Addison-Wesley.

Bradlow, E.T. and Zaslavsky, A. M. (1999), "Analysis of Ordinal Survey Data with No Answer," to appear in the *Journal of the American Statistical Association*.

Bradlow, E.T., Wainer, H., and Wang, X. (1999), "A Bayesian Random Effects Model for Testlets," *Psychometrika*, to appear.

Casella, G. and George, E. (1992), "Explaining the Gibbs Sampler," *American Statistician*, **46**, 167-174.

Caulkin, J., Larkey, P., and Wei, J. (1996), "Adjusting GPA to Reflect Course Difficulty," Working paper, Heinz School of Public Policy and Management, Carnegie Mellon University.

Chaloner, K. (1991), "Bayesian Residual Analysis in the Presence of Censoring," *Biometrika*, **78**, 637-644.

Chaloner, K. and Brant, R. (1988), "A Bayesian Approach to Outlier Detection and Residual Analysis," *Biometrika*, **75**, 651-659.

Chib, S. (1995), "Marginal Likelihood from Gibbs Output," *Journal of the American Statistical Association*, **90**, 1313-1321.

Chib, S. and Greenberg, E. (1995), "Understanding the Metropolis-Hastings Algorithm," *American Statistician*, **49**, 327-335.

Cohen, J. (1960), "A Coefficient of Agreement for Nominal Tables," *Educational and Psychological Measurement*, **20**, 37-46.

Collett, D. (1991), *Modelling Binary Data*, London: Chapman and Hall.

Cowles, M.K. (1996), "Accelerating Monte Carlo Markov Chain Convergence for Cumulative-link Generalized Linear Models," *Statistics and Computing*, **6**, 101-111.

Cowles, M.K., Carlin, B.P., and Connett, J.E. (1996), "Bayesian Tobit Modeling of Longitudinal Ordinal Clinical Trial Compliance Data With Nonignorable Missingness," *Journal of the American Statistical Association*, **91**, 86-98.

Cox, D.R. (1972), "Regression Models and Life-tables" (with discussion), *Journal of the Royal Statistical Society, series B*, **34**, 187-220.

Darwin, C. (1876), *The Effect of Cross- and Self-fertilization in the Vegetable Kingdom*, 2nd edition, London: John Murray.

Dellaportas, P. and Smith, A.F.M. (1993), "Bayesian Inference for Generalized Linear and Proportional Hazards Models via Gibbs Sampling," *Applied Statistics*, **42**, 443-459

DiCiccio, T.J., Kass, R.E., Raftery, A.E., and Wasserman, L. (1997), "Computing Bayes Factors by Combining Simulation and Asymptotic Approximations," *Journal of the American Statistical Association*, **92**, 903-915.

Dobson, A. (1990), *An Introduction to Generalized Linear Models*, London: Chapman and Hall.

Dorn, H.F. (1954), "The Relationship of Cancer of the Lung and the Use of Tobacco," *The American Statistician*, **8**, 7-13.

Elliott, R. and Strenta A. (1988), "Effects of Improving the Reliability of the GPA on Prediction Generally and on Comparative Predictions for Gender and Race Particularly," *Journal of Educational Measurement*, **25**, 333-347.

Fahrmeir, L. and Tutz, G. (1994), *Multivariate Statistical Modeling Based on Generalized Linear Models*, New York: Springer-Verlag.

Fleiss, J.L. (1971), "Measuring Nominal Scale Agreement Among Many Raters," *Psychological Bulletin*, **88**, 322-328.

Gelfand, A.E. (1996), "Model Determination Using Sampling-based Methods," in *Markov Chain Monte Carlo in Practice*, editors Gilks, Richardson, and Spiegelhalter, London: Chapman & Hall, 145-161.

Gelfand, A.E. and Smith, A.F.M. (1990), "Sampling-Based Approaches to Calculating Marginal Densities," *Journal of the American Statistical Association*, **85**, 398-409.

Gelfand, A.E., Hills, S.E., Racine-Poon, A., and Smith, A.F.M. (1990), "Illustration of Bayesian Inference in Normal Data Models Using Gibbs Sampling," *Journal of the American Statistical Association*, **85**, 972-985.

Gelman, A. and Meng, X.L. (1998), "Simulating Normalizing Constants: From Importance Sampling to Bridge Sampling to Path Sampling," *Statistical Science*, **13**, 163-185.

Gelman, A., Carlin, J.B., Stern, H.S., and Rubin, D.B. (1995), *Bayesian Data Analysis*, London: Chapman & Hall.

Gilchrist, W. (1984), *Statistical Modeling,* Chichester: John Wiley and Sons.

Gilks, W., Richardson, S., and Speigelhalter, D.J., editors, (1997), *Markov Chain Monte Carlo in Practice*, London: Chapman & Hall.

Goldman, R. and Widawski, M. (1976), "A Within-Subjects Technique for Comparing College Grading Standards: Implications in the Validity of the Evaluation of College Achievement," *Educational and Psychological Measurement*, **36**, 381-390.

Goldman, R., Schmidt, D., Hewitt, B. and Fisher, R. (1974), "Grading Practices in Different Major Fields," *American Education Research Journal*, **11**, 343-357.

Grayson, D.K. (1990), "Donner Party Deaths: A Demographic Assessment," *Journal of Anthropological Assessment*, **46**, 223-242.

Hambleton, R. and Swaminathan, H. (1985), *Item Response Theory: Principles and Applications*, Boston: Kluwer.

Hand, D.J., Daly, F., Lunn, A.D., McConway, K.J., and Ostrowski, E. (1994), *A Handbook of Small Data Sets*, London: Chapman & Hall.

Hastings, W.K. (1970), "Monte Carlo Sampling Methods Using Markov Chains and Their Applications," *Biometrika*, **57**, 97-109.

Jansen, J. (1991), "Fitting Regression Models to Ordinal Data," *Biometrical Journal*, **33**, 807-815.

Johnson, V.E. (1996), "On Bayesian Analysis of Multirater Ordinal Data," *Journal of the American Statistical Association*, **91**, 42-51.

Johnson, V.E. (1997), "An Alternative to Traditional GPA for Evaluating Student Performance," *Statistical Science*, **12**, 251-278.

Johnson, V.E. (1998), "Posterior Distributions on Normalizing Constants," ISDS Discussion Paper 98-26, Duke University.

Jolayemi, E.T. (1990a), "On the Measurement of Agreement Between Two Raters," *Biometrics Journal*, **32**, 87-93.

Jolayemi, E.T. (1990b), "A Multiraters Agreement Index for Ordinal Classification," *Biometrics Journal*, **33**, 485-492.

Kass, R.E. and Raftery, A.E. (1995), "Bayes Factors," *Journal of the American Statistical Association*, **90**, 773-795.

Landis, J.R. and Koch, G.G. (1977a), "The Measurement of Observer Agreement for Categorical Data," *Biometrics Journal*, 33, 159-174.

Landis, J.R. and Koch, G.G. (1977b), "An Application of Hierarchical Kappa-Type Statistics in the Assessment of Majority Agreement Among Multiple Observers," *Biometrics Journal*, **33**, 363-374.

Larkey, P. and Caulkin J. (1992), "Incentives to Fail," Working paper 92-51, Heinz School, Carnegie Mellon University.

Lee, P. (1989), *Bayesian Statistics: An Introduction*, New York: Oxford University Press.

Li, J., Jaszczak, R.J., Turkington, T.G., Metz, C.E., Musante, D.B., and Coleman, R.E. (1992), "ROC evaluation of Lesion Detectability of Cone Beam versus Parallel Beam Collimation in SPECT with a 3D Brain Phantom," *Radiology*, **185**, 251.

Light, R.J. (1971), "Measures of Response Agreement for Qualitative Data: Some Generalizations and Alternatives," *Psychological Bulletin*, **5**, 365-377.

Lord, F.M. and Novick, M.R. (1968), *Statistical Theories of Mental Test Scores*, Reading, MA: Addison-Wesley.

Manly, B.F.J. (1991), *Randomization and Monte Carlo Methods in Biology*, London: Chapman & Hall.

Masters, G.N. (1982), "A Rasch Model for Partial Credit Scoring," *Psychometrika*, 149-174.

McCullagh, P. (1980), "Regression Models for Ordinal Data," *Journal of the Royal Statistical Society, series B*, **42**, 109-142.

McCullagh, P. and Nelder, J.A., (1989), *Generalized Linear Models*, 2nd edition, London: Chapman & Hall.

McNeil, B.J. and Hanley, J.A. (1984), "Statistical Approaches to the Analysis of Receiver Operator Characteristic (ROC) curves," *Medical Decision Making*, **4**, 137-150.

Mellenberg, G.J. (1994), "Generalized Linear Item Response Theory," *Psychological Bulletin*, **115**, 300-307.

Meng, X.L. and Wong, W.H. (1996), "Simulating Ratios of Normalizing Constants Via a Simple Identity: A Theoretical Exploration," *Statistica Sinica*, **4**, 831-860.

Metz, C.E. (1986), "ROC Methodology in Radiologic Imaging," *Investigative Radiology*, **21**, 720-733.

Nandram, B. and Chen, M.H. (1996), "Reparameterizing the Generalized Linear Model to Accelerate Gibbs Sampler Convergence," *Journal of Statistical Computation and Simulation*, **54**, 129-144.

Nelder, J.A. and Wedderburn, R.W.N. (1972), "Generalized Linear Models," *Journal of the Royal Statistical Society, series A*, **135**, 370-384.

Page, E. (1994), "New Computer Grading of Student Prose, Using Modern Concepts and Software," *Journal of Experimental Education* **62(2)**, 127-142.

Patz, R.J. and Junker, B.W. (1997), " A Straightforward Approach to Markov Chain Monte Carlo Methods for Item Response Models," unpublished manuscript.

Pierce, D.A. and Schafer, D.W. (1986), "Residuals in Generalized Linear Models," *Journal of the American Statistical Association*, **81**, 977-986.

Ramsey F.L. and Schafer, D.W. (1997), *The Statistical Sleuth*, Belmont CA: Duxbury Press.

Rao, C.R. (1973), *Linear Statistical Inference and Applications*, New York: John Wiley & Sons.

Robert, C.P. (1994), *The Bayesian Choice: A Decision-Theoretic Motivation*, New York: Springer-Verlag.

Sahu, S.K. (1997), "Bayesian Estimation and Model Choice in the Item Response Models," Technical report, School of Mathematics, University of Wales, Cardiff, UK.

Samejima, F. (1969), *Estimation of Latent Ability Using a Pattern of Graded Scores,* Psychometrika, Monograph Supplement No. 17.

Samejima, F. (1972), "A General Model for Free Response Data," *Psychometrika*, **38**, 221-233.

Silvapulle, M.J. (1981), "On the Existence of Maximum Likelihood Estimators for the Binomial Response Models," *Journal of the Royal Statistical Society, series B*, **43**, 310-313.

Stout, W.F. (1990), "A New Item Response Theory Modeling Approach With Applications to Unidimensionality Assessment and Ability Estimation," *Psychometrika*, **55**, 293-325.

Strenta, A. and Elliott, R. (1987), "Differential Grading Standards Revisited," *Journal of Educational Measurement*, **24**, 281-291.

Swaminathan, H. and Gifford, J.A. (1982), "Bayesian Estimation in the Rasch Model," *Journal of Educational Statistics*, **7**, 175-192.

Swaminathan, H. and Gifford, J.A. (1985), "Bayesian Estimation in the Two-Parameter Logistic Model," *Psychometrika*, **50**, 349-364.

Swets, J.A. (1979), "ROC Analysis Applied to the Evaluation of Medical Imaging Techniques," *Investigative Radiology*, **14**, 109-121.

Tanner, M., and Wong, W.H. (1987), "The Calculation of Posterior Distributions by Data Augmentation" (with discussion), *Journal of the American Statistical Association,* **82**, 528-550.

Tanner, M. and Young, M. (1985), "Modeling Agreement Among Raters," *Journal of the American Statistical Association*, **80**, 175-180.

Tierney, L. (1994), "Markov Chains for Exploring Posterior Distributions," *Annals of Statistics*, **22**, 1701-1762.

Thissen, D. and Steinberg, L. (1984), "Taxonomy of Item Response Models," *Psychometrika*, **51**, 567-578.

Tsutakawa, R.K. and Lin, H.Y. (1986), "Bayesian Estimation of Item Response Curves," *Psychometrika*, **51**, 251-267.

Tsutakawa, R.K., Shoop, G.L., and Marienfeld, C.J. (1985), "Empirical Bayes Estimation of Cancer Mortality Rates," *Statistics in Medicine*, **4**, 201-212.

Uebersax, J. (1992), "A Review of Modeling Approaches for the Analysis of Observer Agreement," *Investigative Radiology*, **17**, 738-743.

Uebersax, J. (1993), "Statistical Modeling of Expert Ratings on Medical Treatment Appropriateness," *Journal of the American Statistical Association*, **88**, 421-427.

Uebersax, J. and Grove, W.M. (1993), "A Latent Trait Finite Mixture Model for the Analysis of Rating Agreement," *Biometrics*, **49**, 823-835.

van der Linden, W.J. and Hambleton, R.K., editors (1997), *Handbook of Modern Item Response Theory*, New York: Springer-Verlag.

Walser, P. (1969), "Untersuchung uber die Verteilung der Geburtstermine bei der mehrgebarenden Frau," *Helvetica Paediatrica Acta*, Suppl. XX (3), **24**, 1-30.

Young, J.W. (1989), "Developing a Universal Scale for Grades: Investigating Predictive Validity in College Admissions," PhD Thesis, Stanford University, Palo Alto, CA.

Young, J.W. (1990), "Adjusting the Cumulative GPA Using Item Response Theory," *Journal of Educational Measurement*, **12**, 175-186.

Young, J.W. (1992), "A General Linear Model Approach to Adjusting Cumulative GPA," *Journal of Research in Education*, **2**, 31-37.

Young, J.W. (1993), "Grade Adjustment Methods," *Review of Educational Research*, **63**, 151-165.

Index